INFORMATION AND MANAGEMENT SYSTEMS FOR PRODUCT CUSTOMIZATION

INTEGRATED SERIES IN INFORMATION SYSTEMS

Series Editors

Professor Ramesh Sharda
Oklahoma State University

Prof. Dr. Stefan Voß
Universität Hamburg

Other published titles in the series:

E-BUSINESS MANAGEMENT: *Integration of Web Technologies with Business Models*/ Michael J. Shaw

VIRTUAL CORPORATE UNIVERSITIES: *A Matrix of Knowledge and Learning for the New Digital Dawn*/ Walter R.J. Baets & Gert Van der Linden

SCALABLE ENTERPRISE SYSTEMS: *An Introduction to Recent Advances*/ edited by Vittal Prabhu, Soundar Kumara, Manjunath Kamath

LEGAL PROGRAMMING*: Legal Compliance for RFID and Software Agent Ecosystems in Retail Processes and Beyond*/ Brian Subirana and Malcolm Bain

LOGICAL DATA MODELING: *What It Is and How To Do It*/ Alan Chmura and J. Mark Heumann

DESIGNING AND EVALUATING E-MANAGEMENT DECISION TOOLS: *The Integration of Decision and Negotiation Models into Internet-Multimedia Technologies*/ Giampiero E.G. Beroggi

INFORMATION AND MANAGEMENT SYSTEMS FOR PRODUCT CUSTOMIZATION

Thorsten Blecker
Gerhard Friedrich
Bernd Kaluza
Nizar Abdelkafi
Gerold Kreutler
University of Klagenfurt

 Springer

Thorsten Blecker
Gerhard Friedrich
Bernd Kaluza
Nizar Abdelkafi
Gerold Kreutler
University of Klagenfurt, Austria

Library of Congress Cataloging-in-Publication Data
A C.I.P. Catalogue record for this book is available
from the Library of Congress.

ISBN 0-387-23347-4   e-ISBN 0-387-23348-2   Printed on acid-free paper.

Printed in the United States of America.

9 8 7 6 5 4 3 2 1   SPIN 11330011

springeronline.com

Contents

Table of Figures   ix

Table of Symbols   xv

Preface   xxi

Chapter 1: Introduction   **1**

**PART I: FUNDAMENTALS OF PRODUCT
    CUSTOMIZATION**

Chapter 2: Product Customization: Theoretical Basics   **9**

  1.   Product Customization: Definition   10

  2.   Product Customization and Competitive Advantage   11

  3.   Product Customization Strategies   12

  4.   Necessary Conditions when Achieving Product Customization 23

  5.   The Challenging Approach of Product Customization:
       Mass Customization   40

  6.   Summary   43

Chapter 3: Mass Customization and Complexity   **45**

    1.   Complexity: A Literature Review   46

    2.   A System View for Mass Customizing Enterprises   49

    3.   Increasing Complexity Due to Mass Customization   51

    4.   Decreasing Complexity Due to Mass Customization   55

    5.   Interdependencies between Mass Customization and
        Complexity   59

    6.   Summary   61

Chapter 4: A Customers' Needs Model for Mass Customization   **63**

    1.   The Customers' Needs Model   63

    2.   Approaches to Optimize Variety Using the Customers' Needs
        Model   72

    3.   Summary   73

**PART II: AN IT INFRASTRUCTURE FOR EFFECTIVE
AND EFFICIENT PRODUCT CUSTOMIZATION**

Chapter 5: Customer Oriented Interaction Systems   **79**

    1.   Configuration Systems: State of the Art   80

    2.   Advisory Systems as a Means of Customer Support   93

    3.   Extension of the Configuration System with an Advisory
        Component   102

    4.   Summary   113

**Chapter 6: A Multi-Agent System for Coping with Variety
Induced Complexity**   **115**

    1.   Basics of Multi-Agent Systems   117

2. A Multi-Agent Based Approach for Variety Formation and Steering   127

3. Summary   146

Chapter 7: Implementation Scenarios of the Information Systems 149

1. Scenarios and Conditions for a Successful Implementation of the Information Systems   150

2. Summary   160

PART III: CONCEPTS FOR IMPLEMENTING AN EFFICIENT PRODUCT CUSTOMIZATION

Chapter 8: Product Modularity in Mass Customization   163

1. Product Modularity   164

2. Benefits and Limits of Modularity   168

3. Managerial Implications of Modularity   170

4. Selected Tools for the Implementation of Product Modularity   174

5. Summary   179

Chapter 9: Key Metrics System Based Management Tool for Variety Steering and Complexity Evaluation   181

1. Insufficiencies of Current Decision-Supporting Methods in Coping with Complexity   182

2. Key Metrics Systems as a suitable Management Tool   184

3. A Sub-Process Model for the Determination of the Key Metrics 186

4. Complexity Key Metrics   193

5. Extension of the Variety-Sensitive Key Metrics System for Mass Customization   227

6. Summary   241

Chapter 10: Conclusions **243**

References 249

Index 263

Authors 267

Table of Figures

Figure 2-1.   The four approaches to mass customization as a
                response to the customers' sacrifice by Pine/Gilmore   13

Figure 2-2.   Matrix grouping of mass customization configurations
                by Duray et al.   15

Figure 2-3.   Mass customization strategies by Piller   17

Figure 2-4.   Generic levels of mass customization   18

Figure 2-5.   Mass customization modes by MacCarthy et al.   20

Figure 2-6.   Comparison between the classification models for
                mass customization   21

Figure 2-7.   Market-turbulence factors by Pine   24

Figure 2-8.   External and internal conditions necessary for success
                by Kotha   25

Figure 2-9.   Success factors for achieving mass customization
                by Broekhuizen/Alsem   29

Figure 2-10. Necessary conditions for achieving mass customization   31

Figure 2-11. Dimensions of Customer Customization Sensitivity   32

Figure 2-12. Customizing positions on the value chain   34

Figure 2-13. The effect of reducing setup costs on economic
                order quantities   38

Figure 2-14. Matching supply chains with products   39

Figure 3-1.   Complexity of systems   47

Figure 3-2.   Approach of Bliss for the management of complexity   48

Figure 3-3.   The system to be optimized in mass customization   51

Figure 3-4.   Increasing complexity as a consequence of
                variant-rich production   52

Figure 3-5.   Inverted learning curve with variety doubling   53

Figure 3-6.   The enterprise system from a traditional view and
                a mass customization optimal view   55

Figure 3-7.   Mass customization implies modularity   58

Figure 3-8.   Framework presenting the interdependencies between
                mass customization and complexity   60

Figure 4-1.   Information supply and need model   64

Figure 4-2.   Required analogies to establish the customers' needs
                model   65

Figure 4-3.   The objective and subjective customers' needs model   66

Figure 4-4.   Reasons explaining the discrepancies between the
                objective needs, subjective needs and offered variety 68

Figure 4-5.   Identified regions because of the discrepancies between
                the objective and the subjective customers' needs   70

Figure 4-6.   Variety optimization with regard to the objective and
                subjective customers' needs   73

Figure 4-7.   Implications on complexity according to the objective
                and subjective customers' needs model   75

Figure 5-1. Morphological box: classification of configurators 91

Figure 5-2. The Customer Buying Cycle 96

Figure 5-3. Content based and collaborative filtering 100

Figure 5-4. Advisory system architecture 103

Figure 5-5. Conceptual model for advisory processes 105

Figure 5-6. Knowledge acquisition component for the advisory process 106

Figure 5-7. Levers and supporting technologies and tools for advisory system extension 108

Figure 5-8. Extended advisory system structure for mass customization 112

Figure 6-1. Categories of definitions of Artificial Intelligence 118

Figure 6-2. Categorization of Distributed Artificial Intelligence 119

Figure 6-3. An agent in its environment 120

Figure 6-4. Utility-based agents 122

Figure 6-5. A view of a multi-agent system 124

Figure 6-6. Blackboard architecture 125

Figure 6-7. The advisory process at a high level of abstraction 128

Figure 6-8. Framework supporting the multi-agent approach for variety formation and steering 132

Figure 6-9. Influence of product variants' successes and failures on the agent's account 138

Figure 6-10. Module agent's long-term strategy 140

Figure 6-11. Module agent's short-term strategy 141

Figure 6-12. Overview of the variety formation and steering processes 145

Figure 7-1.   Appropriateness of the information system with respect
                to the levels of internal and external complexity   152

Figure 7-2.   Conceptual architecture of the interaction system coping
                with external complexity   154

Figure 7-3.   Technical architecture for an interaction system coping
                with high external complexity.   156

Figure 7-4.   Conceptual architecture of the interaction system coping
                with external and internal complexity.   157

Figure 8-1.   Modular architecture types   165

Figure 8-2.   A graphical representation of the component swapping
                modularity-combinatorial modularity spectrum   166

Figure 8-3.   Classification of the modules strategy with respect to
                cost savings, flexibility and complexity reduction in
                comparison to other strategies   167

Figure 8-4.   Modular function deployment procedure   174

Figure 8-5.   The Module drivers   176

Figure 8-6.   Portfolio architecting process   177

Figure 9-1.   Correspondence between internal abilities and
                sub-processes   187

Figure 9-2.   Relevant sub-processes in mass customization   189

Figure 9-3.   Example for the calculation of the degree of
                commonality index   197

Figure 9-4.   Early versus late differentiation point   207

Figure 9-5.   Retained key metrics and time pattern evaluation   217

Figure 9-6.   Meaning of correlation in the key metrics system   219

Figure 9-7.   Correlations between development key metrics and
                possible variety   220

Figure 9-8.   Correlations between interaction and perceived variety   221

Figure 9-9.   First part of the variety-sensitive key metrics system   221

Figure 9-10. Correlations between component commonality and
production and purchasing process commonalities   222

Figure 9-11. Correlations between component commonality, work-in-
process inventories, differentiation point and setups   223

Figure 9-12. Correlations between setups, manufacturing lead times,
differentiation position and delivery reliability   224

Figure 9-13. Second part of the variety-sensitive key metrics system   225

Figure 9-14. Variety–sensitive key metrics system for mass
customization   226

Figure 9-16. Conceptual illustration of customer differences   230

Figure 9-17. The conceptual value difference curve   231

Figure 9-18. The Kano's model   233

Figure 9-19. Relationship between perceived variety and positiveness   234

Figure 9-20. Key metrics system taking into account the customer's
perspective   239

Figure 9-21. Extended key metrics system for mass customization   241

Table of Symbols

$(N_{al})_i$	Number of attribute levels per attribute i
Ac	Set of all possible actions available to an agent
Acc_{ij}	Account of module agent MA_{ij}
a_i	Value added at process i
A_n	Auction n
AN_{al}	Average number of attribute levels
$APCTE$	Average platform cycle lifetime efficiency
AR	Abortion rate
A_t	Ideal assembly operation time
$AT_{(cc \rightarrow dp)}$	Average time elapsed from configuration until document preparation for manufacturing
AT_c	Average time for carrying out one change in the product configuration system
Bg	Budget of an agent
BT	Maximal bidding time
$bt_n^{\,v}$	Bidding time for aucion n in plan version v
C	Degree of commonality index
$CCR(\Delta T)$	Customers churn rate at the period of time ΔT

CI	Commonality index
c_i	Cost of setup at process i
$CI^{(C)}$	Component part commonality index
$CI^{(p)}$	Process commonality index
C_j	Total cost (material, labor, and overhead) of jth product
C_k	Coalition k
$CR(\Delta T)$	Complaints rate at the period of time ΔT
CT	Average interaction length of time during the configuration process
CT_i	Time needed from one customer to fulfil one configuration
CUM	Capacity utilization metric
d	Total number of distinct component parts used in all the product structures of a product family
d_l	Average throughput time from beginning of production to sale
DI	Differentiation point index
d_i	Average throughput time from process i to sale
D_{j*}	Demand volume of part item j^*
DR	Delivery time reliability
EU	Expected utility
E_v	Multiple use metric
$FIC(\Delta T)$	Frequency of introducing changes to he configuration system at a period ΔT
f_{opt}	Function returning an action from a given set of possible actions, a set of outcomes, a probability distribution and a utility function.
g_i	Steadily decreasing function used in Dutch auction i; $[0,T] \rightarrow]0,1]$; $g_i(0)=1$
h	A particular path from the item d_j to the end item node through the levels of the product tree by including node d_j and excluding the item node
Ic	Interface complexity metric

IL	Integration level of the product configuration system in the existing information systems
j^*	Index of an internally made item on process p according to the sequence that minimizes the total setup times
k	Index of the nodes on path h
k_i	Number of products using the module M_i
$K_i(t)$	Linear decreasing auction function for a module class i
L	Lead time
L_i	Weight of the module M_i
M	Set of all modules
m	Total number of end products in a product family
MA_{ij}	Module agent associated to module variant MV_{ij}
MCM	Modules commonality metric
M_i	Module i
MC_i	Module class i
MV_{ij}	Module variant in a module class M_i
N	Number of fulfilled configurations
n	Number of processes
n^*_{pd}	Total number of part items produced by process p when production is scheduled in the sequence that minimizes the total setup times
nc	Total number of changes introduced in the configuration system
$NC(\Delta T)$	Number of changes and database updates at a period ΔT
N_{ca}	Total number of customizable attributes
n_d	Total number of internally made items, either a final product or component part
n_h	Total number of paths for d_j within product i
NIP	Number of information systems integrated in the configuration system
n_k	Total number of parent nodes on path h
nm	Number of all modules

N_m	Number of manufactured modules in one average product variant
N_{mt}	Total number of modules required to build up all the product variants
no	Number of all orders
$NOC(T)$	Number of customers at the point in time T
$NOLC(\Delta T)$	Number of lost customers at the period of time ΔT
$NONC(\Delta T)$	Number of new customers at the period of time ΔT
n_p	Total number of processes needed for fabrication and assembly
NP	Number of all information systems
N_p	Number of all parts assembled in modules in the plant
n_{pd}	Total number of part items required by a product family to be produced by process p regardless of scheduling sequences
N_v	Number of product variants required by customers
p	Index of a particular process, either for fabrication or assembly
P	Probability distribution over possible outcomes given the performance of actions
PCM	Parts commonality metric
$PCTE$	Platform cycle lifetime efficiency
P_j	Price of each type of purchased parts or the estimated cost of each internally made component part
p_j	Number of part numbers in model j
P_n^{v}	Price that an agent is willing to pay in auction n in plan version v
PPC	Purchasing Process Commonality (within a part category)
$Profits$	Net amount of monetary units that an agent achieves
q_{hk}	Quantity of operation (either manufacturing or assembly)
R	Constant reflecting the risk willingness of a module agent; $R \in [0,1]$
$R(\Delta T)$	Repurchase rate at ΔT

Revenue	Function determining the amount of money that an agent receives when the product variants are displayed to customers.
Risk	Risk function of a module agent
RR(ΔT)	Return rate at the period of time ΔT
SET^*_{pj*}	Setup time of process p for fabricating or assembling part item j^* according to the sequence minimizing total setup time
SET_{pj}	Setup time of process p for fabricating or assembling part j regardless of the sequence minimizing the total setup time
SI	Setup cost index
S_j	Sales in unit of product variant P_j
SM	Setup metric
T	Period of time that a module agent survives without bidding at any auctions
$T_{(cc \to dp)}$	Time elapsed from the completion of configuration i until document preparation
T_∞	Period of time that an agent strives to survive
T_a	Average assembly time for one part
T_{ci}	Time required for change i in the product configuration system
t_i	Average time needed for a setup at process i
T_i	Average assembly time for interfaces between modules
T_{ij}	Task of module agent MA_{ij}
T_j	Average total lead time needed for the manufacturing of jth product
T_{nva}	Average assembly time for non value adding activities
T_t	Average time for functional testing of modules
u	Number of unique part numbers
U	Utility function that associates an outcome with a real number; U: $\Omega \to IR$
UVM	Used variety metric
V_i	Volume of end product i in the family

WIP	Work-in-process turnover
α	Possible action for an agent; $\alpha \in Ac$
β	Total number of immediate parents for all components in the set of end items
ΔT	Period of time
λ_{pj}	0-1 variable / $\lambda_{pj} = 1$ if term j can be made or assembled by process p and $\lambda_{pj} = 0$ otherwise
v_i	Number of different products exciting process i
v_n	Final number of varieties offered
Φ_{ij}	Number of immediate parents for each distinct component part d_j over all the product levels of product i of the family
Φ_j	Number of immediate parents for component j
Ω	Set of all possible outcomes; $\Omega=\{\omega,\omega',..\}$

Preface

Product customization is not a fundamentally new feature in industrial markets. Especially in connection with the mass customization strategy, companies are intensifying their customization processes. Mass customization aims at satisfying the customer's individual needs with near mass production efficiency. In other words, mass customization requires mass production capabilities and make-to-order respectively customization capabilities simultaneously. In essence, mass customization describes the ability of a manufacturing system that produces customized goods in high volume for mass markets by deriving a high number of variants from a single or a few core products.

Although products are still being mass-produced, customers are increasingly demanding an adaptation to their own requirements. For instance, today, there is not one car company that could survive with a narrow range of choices. Mass customization can be applied to a wide range of different products including investment goods, such as machinery, buildings, telecommunication systems, but also consumer goods such as cars, furniture, PCs, and watches. The pursuit of mass customization enables companies to gain an advantage over competitors by providing additional features and benefits.

One of the most critical business success factors in the domain of mass customization is to achieve the flexibility of the product assortment on the one hand, while avoiding cost explosion due to variety and complexity on the other hand. Mass customization affects not only production, but also almost all of the parts and business processes of a company. In addition to the main company functions (source, make, deliver, return) over the supply chain, key enablers such as information systems must be able to master the huge number of possible product variety. Thus, the required flexibility drives high demands on management and technology. Furthermore, best practices in mass customization are rapidly gaining broad attention by managers, who transfer and apply them to all kinds of businesses and industries. At the same time, the academic effort undertaken by researchers of various disciplines to enhance the theoretical foundations of mass customization is constantly increasing.

In this book, the results from a multi-year research project on the "Modeling, Planning, and Assessment of Business Transformation Processes in the Area of Mass Customization" are published. Among other topics, we analyze information and production systems for mass customization and we also develop steering and controlling concepts for the variety induced complexity problem. With the developed tools and frameworks, it is intended to provide basic approaches for coping with the main problems in mass customization, concerning the optimal product assortment to be offered to customers and the product variants to be selected for a particular customer during the interaction process.

"Information and Management Systems for Product Customization" can be positioned at the intersection of Operations Management, Computer Science and Industrial Engineering. It is targeted on a broad audience with both technical and economical backgrounds, especially researchers and students in the mentioned subject areas, but it is also directed towards practitioners in mass customizing industries.

We would like to acknowledge the efforts of Kluwer Academic Publishers, in particular Gary Folven. The research project was funded partly by grants from the Austrian Central Bank, OeNB, No. 9706, and from the Kaerntner Wirtschaftsfoerderungsfonds, KWF.

Chapter 1

INTRODUCTION

Nowadays, the competitive situation of companies is characterized by a very strong orientation towards product individualization. The change from a seller to a buyer market has led to a saturation situation within the industrial goods' markets where the offer by far exceeds the demand. Companies have to struggle for gaining new customers. This major change has increased the customer's power, which has driven companies to differentiate their products from those of competitors by offering individualized problem solutions (Nilles 2002, p. 1). The customer's expectations with respect to services and physical products have also dramatically risen. Therefore, companies tend to increasingly fragment the markets, sometimes to an extreme level, to where each market is occupied by only one customer ("markets of one").

The individualization trend is mainly ascribed to social changes. The high growth of population was a key factor for the emergence of the mass production system, one century ago. But nowadays, especially in the industrial nations, the demographic development shows the population to be steadily decreasing. Simultaneously, wealth and the demand for luxury continuously increase. Psychologists know that in the postmodern era, the need for change and novelty is becoming as important as survival for human beings. The human behavior is essentially determined by the individual principles and is rarely oriented on the behavior of the others (self-determination). It is also well known that if more and more people possess the same object, then the possession of this object is no longer interesting and loses its attractiveness (Piller 1998, p. 22). All of these reasons have contributed to a need for individualization and the demand for products that exactly meet the individual expectations of customers.

Another important trend in the business world is the continuous decreasing of the product life cycles. Consequently, the timeframes for product amortization are considerably reduced. At the same time, the costs of research and development steadily increase because of higher technological complexity of products (Nilles 2002, p. 2). In addition, the ability of fulfilling individual customer needs necessitates the capability of producing a large number of product variants, which induces high costs at both operations and manufacturing-related tasks. In effect, in contrast to the mass production system, in which the economies of scale can be fully utilized, the individualization of customer requirements usually involves a loss of efficiency. On the other hand, globalization and deregulation of markets as well as the rapid diffusion of e-commerce and e-business over the Internet has led to more intensive and aggressive competition. This has also forced companies to develop strategies in order to resist strong price pressures, especially from those companies that are producing in low-wage countries.

The challenge that manufacturing companies have to face is to provide individualized products and services by maintaining a high costs' efficiency. To be successful, companies have to address both of these perspectives, which are necessary for gaining a competitive advantage. The manufacturing of products according to individual customer needs is referred to as product customization. Whereas customization does not necessarily imply a focus on the costs' perspective, in this book we will concentrate on both product customization and costs' efficiency, namely mass customization, which is a new business paradigm that is very challenging for manufacturing companies.

Mass customization is a business strategy that aims at fulfilling individual customer needs with near mass production efficiency (Pine 1993). Whereas the literature includes many contributions that discuss the strategic benefits of mass customization, there are large deficits concerning its implementation in practice. Companies that want to pursue this strategy need a set of practical tools in order to make mass customization work efficiently. The main problem is about how to be able to produce a large number of customer oriented product variants by simultaneously providing prices that do not considerably differ from those of mass products.

Providing customers with individualized products at affordable prices is the main goal of mass customization. However, customers generally accept paying premium prices compared to standard products because they honor the additional benefits of customized products. Therefore, if mass customization fails in providing customers with an optimal or a better solution than any mass products, then the product resulting from the customization process will have, from the customer's perspective, no more additional value than any other standard product. As a result, an optimal understanding of customer needs is a necessary requirement for the success of the strategy. In fact, the focus on customers is not new and not only specific to mass customization. Concepts such as "customer orientation", "close to the customer", "customer segmentation", "niche marketing" and "customer relationship management" reveal the importance of the customer. However, during the pursuit of mass customization, customers have to be seen as partners in the value creation, which implies a deeper customer-supplier relationship. The customer provides a valuable input in the production process and is considered to be a "prosumer" as coined by Toffler (1980, p. 275).

Unfortunately, although the focus on the customer's perspective is a well known issue, the customer in the specific context of mass customization is still misunderstood. Customers are provided with a high number of product variants and are generally supposed to have the capability of making a rational decision. But this is not true because customers are not able to make optimal choices in extensive choice environments. Thus, models for a better conception of customer needs and preferences are required because the customer is a key factor that considerably determines the success or failure of the strategy. Furthermore, a customer orientation through customization tends to trigger increasing costs because of variety and complexity. Suitable tools addressing this relevant issue in the specific case of mass customization are also missed. That is why the intention of this book is to bridge the gap in order to provide information and managerial tools that aim at coping with all of the depicted problems. This book consists of three parts. The first and the second parts each consist of three chapters, whereas the third part has two chapters.

In the first part, we deal with the theoretical basics of mass customization as well as with the main problems that are encountered during the implementation of this strategy. The next chapter (chapter two) can be divided into two main sections. The first section deals with the main configurations of mass customization as described in the literature. This step aims at analyzing state of the art approaches for the practical implementation of mass customization. It is noteworthy that mass customization is a strategy in itself whose implementation can occur according to several strategies. On the basis of this literature review, we discuss the main advantages and disadvantages of each classification of the mass customization strategies. Then, we select the most suitable one reflecting our understanding of mass customization that will be held throughout the entirety of this book. The second section of chapter two deals with the identification of the necessary conditions to be fulfilled for a successful pursuit of the strategy. The main goal is to provide a comprehensive framework that includes the requirements that a manufacturing company has to satisfy prior to moving and while pursuing mass customization. In chapter three, we examine an important issue, which is the complexity problem in mass customization. Because of the extensive variety, a high complexity level is expected in operations and manufacturing related tasks. However, we point out that mass customization has a certain potential for reducing complexity. Based on both of these perspectives of viewing mass customization, we construct a framework that presents the complexity increasing and decreasing aspects of mass customization. Chapter four presents an innovative model explaining the customer requirements in mass customization. This model makes the distinction between the subjective and objective customer needs. The ideas described in this chapter are essential for the development of the concepts and tools that are presented in parts two and three of this book.

Part two deals with the development of appropriate information systems for mass customization. It consists of chapters five, six and seven. Chapter five discusses the state of the art configuration systems in the (web-based) mass customization. Configuration systems can be defined as software tools that enable customers to configure their products online according to their requirements and then to send per mouse-click the product configuration to the mass customizer, which can then begin production. The results attained during the analysis of the existing systems are condensed into a morphological box, which contains all of the dimensions that are up to now considered for the design of configuration systems. A main drawback of these software tools is that they do not strongly consider the customer's perspective. Therefore, we propose to extend configuration systems with advisory systems that are defined as software systems that support customers in finding the products that exactly correspond to their needs and preferences. The technical requirements as well as the infrastructure that is necessary for implementation are described in detail. Chapter six introduces an innovative approach for coping with the internal and external complexity in mass customization. Internal complexity refers to complexity inside the company, whereas external complexity describes the problems that the customer encounters during the decision making process in order to specify the individual product variant. By assuming a modular product architecture, as is common in mass customization, we associate with each module an autonomous rational agent. The defined agents compete with each other in order to form product configurations with the best chances to fulfill customer requirements. Furthermore, the agents provide valuable information about themselves with respect to their suitability of fulfilling the customer requirements and thus their appropriateness to be retained or eliminated from the production program. The multi-agent system simultaneously supports decisions concerning variety formation and steering in mass customization. We also develop the required algorithms that are necessary to make the system work efficiently. The goal of chapter seven is basically to discuss the scenarios, in which the implementation of the advisory system or the multi-agent based approach is suitable. Then, the interfacing possibilities of the configuration system, the advisory system and the multi-agent system will be proposed.

In part three of the book we will deal with product modularity as well as with the development of a managerial tool for the assessment of the variety induced complexity in mass customization. In chapter eight, we focus on modular product structures and their managerial implications. Whereas the information systems developed in part two to a great extent concentrate on the customer's perspective, modularity enables one to put the "mass" in mass customization. It is a very relevant concept that enables the reduction of product complexity and the achievement of the economies of scale, the economies of scope and the economies of substitution by simultaneously ensuring a high level of product variety. Moreover, some practical concepts for the development of modular product concepts will be described. Chapter nine focuses on the development of a managerial tool based on key metrics for controlling the variety induced complexity in mass customization. At first, we identify the main sub-processes in the mass customizing system. Then, we assign to each sub-process the appropriate key metrics. Because single key metrics do not have strong explanatory power, the identified measures are integrated into two key metrics systems. The first key metrics system is called a variety sensitive key metrics system and the second consists of measures that evaluate the effects of the variety-induced complexity in the information sub-process. In order to better take into account the customer's perspective, the variety-sensitive key metrics system is extended. The final system enables one not only to evaluate the internal complexity from the company's perspective, but also the extent of the external complexity perceived by customers when faced with large product assortments. The last chapter ten concludes this book and presents some directions for future research.

PART I

FUNDAMENTALS OF PRODUCT CUSTOMIZATION

Chapter 2

PRODUCT CUSTOMIZATION: THEORETICAL BASICS

Manufacturing companies are all different with respect to the way they meet the market demand. Some companies anticipate the customers' demand and deliver end products from their inventory (make-to-stock). Other companies do not keep finished goods' inventories and manufacture or assemble end products only after receiving a tangible customer order. However, producing finished goods to order does not necessarily mean that the manufactured item is tailored to a specific customer's requirements. For instance, a supplier can choose to produce to order when finished goods are connected with high inventory costs. Furthermore, product customization which involves the supplier's value chain, necessarily assumes the delay of some activities of the value chain until the customer puts in an order.

The trend towards product customization that can be observed nowadays is the result of many changes in the business environment. These have enforced many suppliers to revise their production strategies and management concepts. Many literature contributions emphasize that mass production, as a successful management paradigm, leads to success only under specific conditions. Otherwise, it fails.

Pine (1993, pp. 17) ascribes the extensive development of mass manufacturing in the United States in the early 19th century to the homogeneity of the American market. Furthermore, the individual income was equitable and customers had similar needs and requirements. Mass production provides a mass market with goods at a consistent quality and affordable prices. It builds upon main principals that include among others: economies of scale, product standardization, specialization, division of labor, hierarchical organization, and vertical integration. The main goal is to develop, manufacture, market and deliver goods and services at prices which are low enough to where nearly everyone is able to afford them. Pine (1993, p. 25) speaks about a feedback loop that has characterized the interplay between customers and suppliers. This feedback loop has strengthened standardized products, mass production techniques and large, homogeneous markets. As a result, it was not necessary to offer several product options. For instance, Henry Ford promised his customers to receive any car color they would like to have, as long as it was black.

Mass production was accepted and successfully adopted by many manufacturing enterprises. It builds upon the precepts of the scientific management (Taylorism) and strongly focuses on operational efficiency and productivity. The main objective is to enable mass manufacturers to lower costs and to sell products at affordable prices. Mass production is also favored by a seller market, in which the customers' demand exceeds the offer. In such a market, suppliers are more powerful than customers because they basically instruct their customers what to buy. However, these market conditions have been enormously altered because of several changes that occurred in the economical environment such as input instabilities, changing demographics, changing needs and wants, saturated markets, demand uncertainties, innovations (Pine 1993, p. 32), etc. Thus, the main conditions that have ensured a successful mass production, namely stability and demand homogeneity are no longer available and do not coin the actual picture of the business environment. Nowadays, customers have more power and the suppliers' offer by far exceeds the demand. In contrast to the seller market, the involved market is called a buyer market, in which the mass production paradigm no longer represents a successful solution model. In many business fields, suppliers offer a wide range of choices in order to increase the likelihood that customers find the suitable product. However, the tendency observed is that large variety is not enough because customers want customized products that optimally fulfill their requirements.

1.   PRODUCT CUSTOMIZATION: DEFINITION

Pine/Gilmore (1999, p. 76) define customization as "...producing in response to a particular customer's desires." The authors point out that it is relevant to make the distinction between variety and customization. Whereas customization strives for fulfilling individual customer's needs, variety simply involves more choice from which the customer is able to choose. "Fundamentally, customers do not want choice; they just want exactly what they want" (Pine/Gilmore 1999, p. 76). Customization is intended to add increased customer perceived value to a product, since a customized product

– compared to a mass produced product – increasingly fulfills the need of the customer (Svensson/Jensen 2001, p. 1).

When defining the term product customization, it is relevant to include the product perspective which can be a physical good or a service. Thus, product customization can be defined as producing a physical good or a service that is tailored to a particular customer's requirements. In this context, customer involvement is an important issue, because customers dictate what the enterprise has to produce. In the case of physical goods, product customization can occur ex post after manufacturing by the retailer or the customers themselves. In this book only physical goods' customization is considered because the main focus will be placed on manufacturing enterprises. However, value adding services around physical products are considered as additional criteria for differentiating goods and thus for customization.

2. PRODUCT CUSTOMIZATION AND COMPETITIVE ADVANTAGE

"Competitive advantage fundamentally grows out of the value a firm is able to create for its buyers that exceeds the firm's cost of creating it. Value is what buyers are willing to pay, and superior value stems from offering lower prices than competitors for equivalent benefits or providing unique benefits that more than offset a higher price" (Porter 1998, p. 3). It is obvious that Porter emphasizes the value offered to the customer as the most important factor which determines the extent of competitive advantage. In this context, offering customized products seems to be a source of competitive advantage because the ability to develop customer-tailored products can be marketed as a differentiating and distinctive capability that provides customers with superior value. Therefore, product customization can be considered as a feature to differentiate goods from those of competitors.

In contrast to mass produced goods that are designed for an average customer, customized products respond to particular requirements. A mass manufactured product may consist of product options or features that customers do not need. Thus, customers have to pay for components which are undesired and do not provide an additional value for them. But when the product is customized, the undesired features as in the case of mass production are not available and customers just pay for the product with the configuration that satisfies their requirements (Pine/Gilmore 1999, p. 79). For example, suppose that all variants of a car model of an automobile manufacturer are equipped with a navigation system in which the customers appreciate the value of this device differently. Even when customers do not need this system they have to pay for it. Some customers may value the price of the car to be high and look for competitors' offers, although costs and price would be relatively lowered without such a system. This represents a waste that increases costs from a supplier's perspective. It would be cleverer to offer the navigation system as an option that can be freely selected by customers. Therefore, product customization appears as a way to considerably reduce costs as well as price.

Recapitulating, product customization enables suppliers to consider both basic types of competitive advantage that are identified by Porter (1998), namely the cost and differentiation perspectives. Although when identifying the generic competitive strategies, Porter mentions that firms have to choose either cost leadership or differentiation and all suppliers who are in between are "stuck in the middle", product customization appears as a hybrid competitive strategy that is able to combine both advantages together.

3.   PRODUCT CUSTOMIZATION STRATEGIES

In order to customize products there are two main approaches. The first is to specifically design and manufacture products for a particular customer's requirements by using a job shop manufacturing system. These products are designed and produced from scratch for each individual customer, in other words, the needed resources are used differently to a great extent for each specific product. For instance, building an airport having particular characteristics is a specific project that is constructed once for one customer. However, the second approach is to implement a mass customization (Davis 1987; Pine 1993) system that aims at linking both advantages of mass production and customization. The main objective of mass customization is to produce individualized goods with near mass production efficiency (Pine 1993). The first approach is considered to be a project management problem and will not be dealt with in this book, the second approach of mass customization is very challenging. Thereby, not only the product individualization perspective plays a relevant role, but also the costs' perspective. For example, Rautenstrauch et al. (2002, p. 104) speak about mass customization when the product price does not exceed approximately 10-15 percent of a standard product. From a strategic point of view, the goal of mass customization is to differentiate products through customization and to also take advantage of the economies of scale. Piller (2000, p. 196) mentions that the mass customizer has to provide customers with an achievement potential by developing a wide product solution space from which customers can select or self-configure the product variant that meets individual requirements. The rest of this book will basically concentrate on the mass customization paradigm. To implement mass customization, there are many different strategies which have been already discussed in the technical literature. In the following, the main identified mass customization configurations are presented.

3.1 Mass Customization Strategies by Pine/Gilmore (1999)

Pine/Gilmore (1999, pp. 86) introduced a taxonomy to classify suppliers who pursue mass customization. In their classification, two main dimensions are taken into account, namely the product and its representation. To elaborate the model represented by figure 2-1, the authors introduce the notion of customer sacrifice which is defined as "the difference between what a customer accepts and what he really needs, even if the customer doesn't know what that is or can't articulate it" (Pine/Gilmore 1999, p. 78). In order to respond to the customer sacrifice, the mass customizer can make decisions as to whether to change the product functionalities or not. Moreover, companies have the possibility to change the representation of the product or not, which relates to anything else outside the product itself such as the product description, packaging, name, etc.

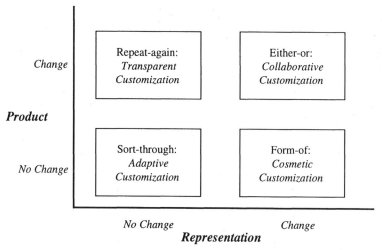

Figure 2-1. The four approaches to mass customization as a response to the customers' sacrifice by Pine/Gilmore (1999, p. 87)

When the product and its representation are changed, it is referred to the corresponding customization strategy as collaborative customization. The main objective is that customers no longer have to make either-or sacrifices. For example, when buying a standard pair of shoes one customer may wish for a certain model but the shoes' size may not fit his or her feet well. However, another pair of shoes may exist in the assortment that exactly fits customer's feet, but the shoes' model design may be undesired by the customer. In this case, the customer has to make an either-or sacrifice. In order to avoid that, a customer renounces one product dimension (either the shoes' model or size), collaborative customizers interact directly with their customers to determine what they want in order to manufacture the product that best fits their requirements.

However, when several alternative customers' requirements can be embedded into one single product, adaptive customization becomes an interesting approach. Thereby, the product as well as its representation remain unchanged. Adaptive customization is considered as a response to a cumbersome sort-through process that customers engage in when presented with too many finished goods. The outcome of this type of customization is adaptive products which are, according to Pine (1993, p. 14), standard products that can be adapted for or by customers themselves. These products can be either customizable or customizing. Whereas customizable products provide customers with the possibility to choose from many options, the one that most suits their specific requirements (e.g. graphic equalizer of a hi-fi system), customizing products adapt themselves to the user (e.g. Gillette Sensor).

In the case when it is only the product representation that has changed, Pine/Gilmore calls the corresponding mode cosmetic customization. The main goal is to avoid form-of sacrifices by differently presenting a standard product to a multitude of customers. For example, a standard product can be specially packaged for a particular customer's requirements.

In contrast to cosmetic customization, transparent customization assumes that the representation of the product does not change, whereas the product functionalities are adapted to particular requirements. This type of customization aims at eliminating repeat-again sacrifices that customers have to encounter each time they have to perform the same task of specifying their requirements again and in turn providing them to the supplier. Thus, transparent customization is suitable when customers do not want to be bothered with direct collaboration. Pine/Gilmore (1999, pp. 92-93) give the example of Chemstation, an industrial cleaning goods' manufacturer and distributor that provides its customers with individualized goods that suit their particular facilities without explicitly letting them know that the product is customized. In this case, transparent customization is an advantageous approach because customers rather focus on the cleanness of their facilities than on the attributes or the chemical composition of the cleaning product.

3.2   Mass Customization Strategies by Duray et al. (2000)

The model of Pine/Gilmore is customer oriented and basically classifies mass customization according to its capability of avoiding a specific sacrifice that the customers have to make each time they settle for products that do not exactly correspond to their needs. However, the classification described by Duray et al. (2000, pp. 612) rather focuses on an operations perspective. In their model (Figure 2-2), Duray et al. present a taxonomy of mass customization with respect to two dimensions, namely the point of customer involvement and the type of modularity. These dimensions are analyzed in relation to the production cycle consisting of the design, fabrication, assembly and use phases.

Type of Modularity

Point of Customer Involvement	Design	Fabrication	Assembly	Use
Design				
	1 Fabricators		2 Involvers	
Fabrication				
Assembly				
	3 Modularizers		4 Assemblers	
Use				

Figure 2-2. Matrix grouping of mass customization configurations by Duray et al. (2000, p. 612)

The customer involvement dimension is considered to be an indicator of the product customization level. If customers are involved in the design and fabrication stages, then the degree of customization is high. However, when customers are involved in the assembly or use phases customer involvement is considered to be lower. The second dimension which refers to the type of modularity is essential because it enables companies to put the "mass" in mass customization and to reach lower costs when manufacturing customized products. Duray et al. make the distinction between modularity allowing the modification of modules and components during design or fabrication and modularity that only involves unchangeable standard modules in the assembly and use stages.

By juxtaposing both dimensions, Duray et al. create four types of mass customizers which are fabricators, involvers, modularizers and assemblers. Fabricators involve customers early on in the production cycle and use modularity to increase commonality of components. New modules can be developed or already existing ones can be modified. In opposition to fabricators, involvers use modularity in the late stages of the production cycle. Due to early customer involvement, involvers give the impression to customers that the product is specifically designed for them. However, customer requirements are met by assembling or using standard modules and components.

Modularizers are manufacturers that use modularization at the first stages of the production cycle but involve customers during assembly and use. At a first glance, this approach may appear inconsistent, especially for those who have a picture in mind that the single goal of module combinations is to suit a specific customer configuration. However, modularizers are companies who use modularity in the first stages in order to increase internal commonality between products. *"Modularizers* incorporate both customizable modularity in the later stages of the production cycle and non-customizable modularity in the design and fabrication stages of the production cycle" (Duray et al. 1999, p. 613).

Finally, assemblers are mass customizers that pursue an assemble-to-order strategy. They involve customers in the late stages of the production cycle and manufacture customized products on the basis of standard modules.

3.3   Mass Customization Strategies by Piller (2000)

In order to classify mass customization, Piller (2000, pp. 250) makes the distinction between soft and hard customization (Figure 2-3). Soft customization is solely based on activities of research and development, design engineering or sales. It involves only a few product variants that are manufactured in large batches. The soft customized products can either be individualized by the customers themselves (self customization) or by retailers (point-of-delivery customization). Soft customization can also result from secondary services around a standard product (service customization) in order to provide the impression that the product is tailored to individual customers' requirements. Thus, soft customization basically builds upon a mass product with standardized manufacturing processes. In other words, customers have no direct influence on the production process.

Soft Customization: Customization based on fully standardized manufacturing processes	*Hard Customization*: Customization starts within the manufacturing processes
Self customization Create customizable products and services	**Customization-Standardization-mix** Either the first or the last activities of the value chain are customized within the factory, while keeping the others standardized
Point-of-delivery customization Customization of a standardized product at the point-of-delivery	**Modular product architectures** Modularize components and combine them to customized products
Service customization Customize services around standardized products and services	**Flexible customization** Using flexible manufacturing systems for production of fully customized products without higher costs

Figure 2-3. Mass customization strategies by Piller (2000, p. 251)

In opposition, hard customization starts within the manufacturing processes because some production activities will not be achieved until customers specify their requirements. Each manufactured product can be unambiguously attributed to one single customer order. Three main modes can be discerned which are: customization-standardization mix, modular product architectures and flexible customization. With the customization-standardization mix, Piller (2000, pp. 256) considers product individualization where customers are integrated either at the first or last stages of the value chain. This mode of customization can be based on modularity whereby a set of standard parts are combined with a single individualized component, which can occur at the beginning or at the end of the manufacturing process. The customization mode that is referred to as modular product architectures involves a more advanced modularity where product variants can be configured on the basis of modular building blocks according to individual requirements. Whereas modular architectures considerably restrain the number of choices that are available to customers, flexible customization essentially is based on flexible manufacturing systems in order to produce a batch size of one by keeping costs as low as possible. Thereby, the individualized components represent the main elements of the product. This mode of mass customization has to be distinguished from job-shop manufacturing where costs are very high and not comparable to those of corresponding standard products. Furthermore, manufacturing processes should be stable and mechanisms have to be established in order to ensure an efficient management of customer-oriented designs.

3.4   Mass Customization Strategies by Da Silveira et al. (2001)

On the basis of a literature analysis, Da Silveira et al. (2001, pp. 3-4) introduce a classification framework for mass customization levels. The established framework consists of eight generic levels and ranges from pure standardization to pure customization (Figure 2-4).

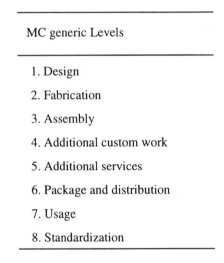

MC generic Levels
1. Design
2. Fabrication
3. Assembly
4. Additional custom work
5. Additional services
6. Package and distribution
7. Usage
8. Standardization

Figure 2-4. Generic levels of mass customization
(Adapted from: Da Silveira et al. 2000, p. 3)

Design refers to a collaborative project in the sense of Pine/Gilmore (1999) where customers interact with suppliers with the objective to design a product that fulfills particular requirements. However, *fabrication* is targeted on the manufacture of customized products by following basic predefined designs. Whereas fabrication allows the introduction of some modifications into the product building blocks, *assembly* is based on standard modules that can be combined into different product variants in order to meet particular customer requirements. In opposition to the first three levels: design, fabrication and assembly where the customer actively intervenes in the manufacturing process, in the five other levels customers have no influence on production. In levels 4 and 5, mass customization is achieved by providing *additional custom work* at the point of delivery or secondary *additional services* around standard products. The *package and distribution* level involves a cosmetic customization. Finally, *usage* refers to customizable products, whereas *standardization* relates to customizing products which self adapts to specific customer needs (Da Silveira et al. 2001, pp. 3-4).

3.5 Mass Customization Strategies by MacCarthy et al. (2003)

Based on a literature analysis and a case studies' approach, MacCarthy et al. (2003, pp. 290) determine three dimensions according to which they define five modes for mass customization.

The first dimension refers to product design and product validation that can be achieved per product family, per product or per order. In the case of a product family, design and validation processes are completely achieved before any customer places an order. However, in the per product situation the supplier involves a particular customer in the design and validation processes before order placement. Then, the customized product can be ordered on a repeat basis without customer involvement. In contrast, the per order case assumes that the customer is involved at any time he or she places an order.

The second dimension relates to the resources utilized for order fulfillment. These resources can be either fixed or modifiable. Mass customizers with fixed resources predefine their customization capabilities, so that customers' orders are fulfilled on the basis of e.g. processing or delivery resources that are predetermined in advance. However, modifiable resources offer a larger degree of flexibility because mass customizers can engage suppliers with specific capabilities in order to suit particular customers' requirements.

The third dimension detects whether the product is based on a once-only or call-off approach. Whereas once-only means that the product is manufactured only once for one customer and that the probability of receiving identical orders is very low, the call-off approach is used in the business-to-business field when customers are mass merchants who can request customized orders in large batch sizes.

Although the juxtaposition of all three dimensions would theoretically lead to 3x2x2= 12 modes of mass customization, MacCarthy et al. (2003, p. 297) argue that some configurations are infeasible. They retain only five fundamental modes which are catalog mass customization (mode A), fixed resource design-per-order mass customization (mode B), flexible resource design-per-order mass customization (mode C), fixed resource call-off mass customization (mode D), and flexible resource call-off mass customization (mode E). The following figure 2-5 indicates for each fundamental mode of mass customization the corresponding combination of the dimension values.

	A	B	C	D	E
	Catalogue	Fixed resource design-per-order MC	Flexible resource design-per-order MC	Fixed resource call-off MC	Flexible resource call-off MC
Product design	Per family	Per order	Per order	Per product	Per product
Product validation	Per family	Per order	Per order	Per product	Per product
Once-only/call-off	--	Once-only	Once-only	Call-off	Call-off
Fixed/modifiable resources	Fixed	Fixed	Modifiable	Fixed	Modifiable

(c) Elsevier 2003, Reproduced with permission.

Figure 2-5. Mass customization modes by MacCarthy et al. (2003, p. 299)

3.6   Critical Comparison Between the Classification Models for Mass Customization

In order to evaluate the classification models for mass customization, a comparison based on relevant criteria has to be carried out. We identify five basic dimensions according to which we compare the different models, namely: research type, exclusiveness between strategies, main classification perspective, easiness of attribution and specification of application suitability. *Research type* indicates the research method on which the basis of the model is established. The dimension: *exclusiveness between strategies* captures whether the strategies in one model are mutually exclusive, in other words, if there is some overlapping in the classification. The *main classification perspective* refers to the basic criterion that is used in order to achieve the typology (e.g. the classification perspective according to Pine/Gilmore (1999) is: product/representation). The *easiness of attribution* relates to the fact of whether difficulties arise when trying to assign a particular mass customizer to one category according to the model. The *specification of application suitability* means whether the researchers have specified in their original contributions the context, in which their classification makes sense. For example, when considering an operations-oriented problem a particular classification model may be more suitable than another model that can be for instance more applicable when dealing with a marketing-oriented problem. Figure 2-6 summarizes the evaluation of the classification models with respect to the identified dimensions.

	Pine/Gilmore (1999)	Duray et al. (2000)	Piller (2000)	Da Silveira et al. (2001)	MacCarthy et al. (2003)
Research type	Empirical investigation	Empirical research and validation	Literature research and 101 case studies	Literature research	Literature research and 5 case studies
Exclusiveness between strategies	Available	Available	Not available	Available	Available
Main classification perspective	Product/ representation	Point of Customer involvement/ modularity	Customer involvement in value chain	Customer involvement in value chain	Product design/ repetition of orders/ resources
Easiness of attribution	Easy	Not easy	Not easy	Not easy	Easy
Specification of application suitability	None	None	None	None	None

Figure 2-6. Comparison between the classification models for mass customization

The models by Pine/Gilmore (1999) and Duray et al. (2000) are based on empirical researches. Pine/Gilmore (1999) introduce the notion of customer sacrifice and derive their model that classifies mass customization into four main strategies. For each strategy, the authors provide a list of suitable illustrating examples from the practice. But an empirical validation of the model is not available. However, Duray et al. (2000) first theoretically conceive their typology and then validate it empirically on the basis of a survey involving the participation of 126 mass customizers.

In order to conceptualize their models, Piller (2000) and MacCarthy et al. (2003) carry out a literature research as well as a case studies analysis. It is relevant to mention that the case studies analysis achieved by Piller is by far more comprehensive than the analysis made by MacCarthy et al. In contrast, Da Silveira et al. (2001) base their model uniquely on a literature research.

The analysis of the different models with respect to the exclusiveness between strategies reveals that the strategies identified by Piller (2000) are not mutually exclusive. This basically concerns the point-of-delivery customization strategy which appears as a particular case of the customization-standardization mix because point-of-delivery customization can be seen as a form of customization that occurs at the last stage of the value chain. In addition, the customization-standardization mix can be seen as a particular case of modular product architectures.

With respect to the main classification perspective, Duray et al. (2000), Piller (2000) and Da Silveira et al. (2001) choose a value chain perspective, whereas Pine/Gilmore (1999) and MacCarthy et al. (2003) base their models on other dimensions. It is noteworthy that it may not be straightforward to assign mass customizers to a suitable category when applying the approaches built upon a value chain perspective since these approaches are based on an implicit premise assuming a single point of customization (MacCarthy et al. 2003, p. 295). In many practical cases, it can be observed that there are many points of customization across the value chain. Therefore, the models by Pine/Gilmore (1999) and MacCarthy et al. (2003) are the most straightforward to be applied for the assignment of a mass customizer to a particular category.

An important issue that remains generally unaddressed by each described research relates to the application limits of each model. For example, the model by Duray et al. (2000) ignores mass customizers who offer standard products that are customizable by retail or by the customers themselves. The identified configurations: fabricators, involvers, modularizers and assemblers can be assigned to collaborative customization in the sense of Gilmore/Pine (1999) because suppliers usually involve their customers and directly interact with them. Thus, the model by Duray et al. (2000) can be considered as a further classification of collaborative customizers with the main focus on a manufacturing perspective. This model is rather relevant when addressing research issues in operations management for mass customization such as complexity in manufacturing. However, the model of Pine/Gilmore (1999) seems to be more suitable when addressing a customer's perspective.

Although in all cases the application limits of each model are not explicitly mentioned, these can be derived from the different definitions for mass customization adopted in each research. For example, Duray et al. (2000) speak about mass customization only when customers are directly involved in the value chain, either in the early or late stages and when products are modularized. In addition to modularization and customer involvement, Pine/Gilmore (1999) consider customizable products or standard products with tailored services as a form of mass customization. MacCarthy et al. (2003) include suppliers in their classification that manufacture customized products in large batches for mass merchants. Piller (2000) and Da Silveira et al. (2001) base their analysis on a similar definition as Pine/Gilmore (1999) and also consider adaptable products as well as customization through services as a form of customization. However, in contrast to Da Silveira et al. (2001), Piller (2000) explicitly distinguishes between mass customization configurations where customers have a direct influence on the manufacturing process and those configurations where customers are not directly involved in production. A main challenge to be addressed in this work is to provide a definition for mass customization, on which basis we can determine what problems managers encounter during implementation. But before dealing with a definition of this new business paradigm, we will analyze the necessary conditions that must be available in order to shift to and pursue mass customization.

4. NECESSARY CONDITIONS WHEN ACHIEVING PRODUCT CUSTOMIZATION

This section deals with the necessary conditions that must be available when achieving product customization in the case of the mass customization paradigm. The main concern is to determine the specific organizational requirements and favorable market factors that will contribute to an increase in the probability of the success implementation of mass customization. For instance, according to Broekhuizen/Alsem (2002, p. 313) success in mass customization is attained when the supplier provides customers with "...*superior customer value* – in contrast to mass manufacturers' offerings – through customization on a mass scale".

As we will see in the literature review, the terms used by the authors to refer to what we call necessary conditions are quite different. This term is intentionally chosen in order to avoid the use of e.g. the term success factor which is unfortunately, arbitrarily utilized in the technical literature. We argue that the term success factors has to be exclusively used when addressing strategic management issues. That is why, we opt for the definition stated by Kaluza (1989, p. 28) who identifies six strategic success factors, namely: costs, quality, time, flexibility, service, and product variety. However, necessary conditions can relate to success factors, competences or capabilities that the supplier has to examine and to develop further before and during the pursuit of mass customization.

4.1 Literature Review

Pine (1993, pp. 54) develops a market-turbulence map that aims at supporting managers in an evaluation if their organizations need to move into mass customization. "The greater the market turbulence, the more likely that the industry is moving toward mass customization, and that the firm *has* to move in order to remain competitive" (Pine 1993, pp. 54-55). The elaborated market-turbulence map includes 17 factors and is structured in two main categories which are "demand" and "structural" factors (Figure 2-7). Demand factors such as stability of demand levels, uncertain customer needs or heterogeneity of demand basically relate to the customer. They can be manipulated by the firm itself which can stabilize demand by e.g. retreating from the more turbulent segments. However, structural factors such as vulnerability to substitute products, product life cycle length or the rate of product technology change relate to the basic nature of an industry. Therefore, the manipulation of these factors by firms themselves is more difficult. However, the market turbulence introduced by Pine only covers the dimension of market conditions and does not deal with the organizational requirements that are necessary for the implementation of mass customization.

Demand factors

- Stability and predictability of demand levels
- Basic necessities versus complete luxuries
- Easily defined versus uncertain customer needs/wants
- Homogeneous versus heterogeneous demand
- Rate of change in customer needs/wants
- Price consciousness
- Quality consciousness
- Fashion/style consciousness
- Level of pre- and postsale service

Structural factors

- Buyer power
- Degree of influence of economic cycles
- Competitive intensity
- Price competition versus product differentiation
- Level of market saturation
- Vulnerability to substitute products
- Product life cycle length and predictability
- Rate of product technology change

Figure 2-7. Market-turbulence factors by Pine
(Adapted from: Pine 1993, pp. 56)

Pine (1993) considers the shift to mass customization as an inevitable reaction when firms want to be successful in turbulent market environments. He postulates that a competitive advantage can be attained if firms completely replace "mass production" with "mass customization". In contrast, Kotha (1995, pp. 25) demonstrates through a case study of the National Industrial Bicycle Company of Japan (NIBC) the compatibility of both strategies. He argues that competitive success in changing environments can be achieved, if firms have the necessary capabilities to create knowledge by interacting mass customization with mass production. Kotha (1996, pp. 447) addresses the basic conditions which are necessary for a successful implementation of mass customization. He makes the distinction between external (industry-level) and internal (firm-level) conditions. Figure 2-8 presents the necessary conditions for a successful pursuit of mass customization as discussed by Kotha (1996, p. 449).

External Conditions

Success is more likely if
- there is no well entrenched competitor already pursuing mass customization
- the firm has access to a supplier network in close proximity
- the industry is characterized by increased product proliferation and new product introductions
- the firm develops an inter-connected information network with a selected group of trained retailers

---

Internal Conditions

Success is more likely when a firm
- has made long-term investment in advanced manufacturing technologies and information technologies and human resource development
- has access to substantial in-house engineering expertise and manufacturing capabilities
- focuses its manufacturing tasks and competitive priorities at each plant to its product/market environment
- institutes organizational mechanisms that foster interactions among focused plants
- creates a culture that emphasizes knowledge creation and the development of manufacturing capabilities
- has a savvy marketing group that can excite customers about individualized product offerings

---

Figure 2-8. External and internal conditions necessary for success by Kotha (1996, p. 449)

A careful examination of the external and internal conditions for success suggests that Kotha has orientated his analysis too much on the case study of the NIBC. He assumes that both mass production and mass customization are pursued in two different focused plants by the firm. The analysis merely concentrates on mass producers who want to additionally pursue mass customization by offering individualized products to a specific customer segment. But the conditions discussed by Kotha will not be helpful for a startup firm that wants to offer mass customized goods. The model also does not consider the specific case when the company pursues a craft production and wants to move into mass customization.

According to Hart (1995, pp. 39), when companies intend to move into mass customization, four key decisive factors have to be examined in order to attain successful implementation: (a) customer customization sensitivity, (b) process amenability, (c) competitive environment, and (d) organizational readiness. Customer customization sensitivity is based on two main factors, which are: the uniqueness of customers' needs and customer sacrifice. The uniqueness of customers' needs basically depends on the nature of the product. For example, whereas customers would not really benefit from customized table salt, with respect to investment counseling, each customers' needs are absolutely unique. Customer sacrifice is defined in accordance with Pine/Gilmore (1999, p. 78) as the gap between what a customer wants exactly and what he or she settles for. The higher the level of the needs' uniqueness and customer sacrifice, the higher the customization sensitivity level.

The second key factor: process amenability encompasses the enablers which fall into two categories: technology based and organizational, as well as the following functions: marketing, design, production, and distribution. The main concern should be attributed to the assessment whether the necessary competencies are available for a shift to mass customization. The third factor: competitive environment has to be examined with respect to the existence or absence of competitive forces that would enhance or detract from the advantage gained when implementing mass customization. The fourth factor: organizational readiness aims at discovering the degree of suitability between the business opportunity inherent in mass customization and the organization's ability to efficiently pursue mass customization.

In their literature review on mass customization, Da Silveira et al. (2001, p. 4) discuss six factors in order to successfully achieve mass customization, which are: (a) customer demand for variety and customization, (b) appropriateness of market conditions, (c) readiness of the value chain, (d) availability of technology, (e) customizability of products and (f) knowledge sharing.

The first two factors (a) and (b) are market-related, whereas the other factors are rather organization-based. *Customer demand for variety and customization* is the main impulse that drives suppliers to develop processes and capabilities for product customization. If customers choose to not pay premium prices or to wait for longer delivery times, then mass customization is superfluous and success is improbable. Furthermore, success may depend on the point in time at which the company starts to offer individualized products. Being the first supplier over competitors that pursues mass customization can represent a significant advantage and substantially improves the supplier's image. This is referred to as the *appropriateness of market conditions*.

The achievement of mass customization is a complex task that assumes an optimal coordination and harmonization between all members of the value chain including suppliers, distributors and retailers. The achievement of this goal supposes the *readiness of the value chain* as well as the *availability of technology* such as modern information systems. Modern technology on the shop floor offering a high manufacturing flexibility (Kaluza 1995, p. 15) is also required in order to customize products. Flexibility in manufacturing has to be supported by a modular product design which in turn determines the *customizability of products*. In addition, the design of customized goods supposes a near interaction between suppliers and customers, which requires a comprehensive *knowledge sharing* across the whole value chain.

In the same context, Zipkin (2001, p. 82) defines three key capabilities of mass customization systems, which are: "*elicitation* (a mechanism for interacting with the customer and obtaining specific information); *process flexibility* (production technology that fabricates the product according to the information); and *logistics* (subsequent processing stages and distribution that are able to maintain the identity of each item and to deliver the right one to the right customer)."

Berman (2002, p. 59) develops a checklist for mass customization readiness on the basis of the key capabilities identified by Zipkin (2001). He mentions that suppliers have to compute the financial benefits associated with mass customization. They have to evaluate the extent of sacrifice when customers settle for standard solutions as well as the segment size of customers who would be interested in individualized products. A supplier has to ensure that technical and personal skills for customer needs' elicitation already exist or can be easily developed. Manufacturing capabilities should be assessed with respect to product modularity, postponement of orders and flexible production systems including computer-aided design and computer-aided manufacturing. An efficient achievement of mass customization also depends on efficient communication across the supply chain with suppliers and customers. Furthermore, the use of just-in-time techniques, bar coding and scanning as well as the ability to run mass production at the same time as mass customization must be analyzed before moving into mass customization.

Broekhuizen/Alsem (2002, p. 313) criticize the fact that in the technical literature, there is no coherent and detailed framework according to which success factors for mass customization are classified. That is why, on the basis of a literature analysis, they develop a conceptual model that is structured according to five dimensions which are the customer, product, market, industry and organization. Figure 2-9 summarizes the hypotheses made by the authors with respect to the probability of success in mass customization.

Mchunu et al. (2003, pp. 6) do not speak about success factors or capabilities, but rather about competencies. They point out that there are five competencies with specific relevance for mass customization, which are:

- *Design* that has to be oriented on customers' interests
- Strategic and operational *flexibility*
- *Supply chain agility* with the main focus on market responsiveness
- *Distribution of inventory* with respect to the amount and position of material throughout the company's facilities
- *Logistics and information management*

Building upon these five key competencies, Mchunu et al. (2003, p. 10) propose a framework on which basis they claim capable of predicting the success or failure of suppliers who want to move into mass customization. They apply their predictive framework for two practical cases in order to assess whether the suppliers are fit for adopting a mass customization strategy or not.

Customer factors

H1a: Consumer heterogeneity positively affects the probability of success

H1b: Consumer involvement positively affects the probability of success

H1c: Consumer willingness to pay a price premium positively affects the probability of success

H1d: Consumer privacy concerns negatively affect the probability of success

Product factors

H2a: Purchasing frequency positively affects the probability of success

H2b: Product luxury level positively affects the probability of success

H2c: Product visibility positively affects the probability of success

H2d: Product adaptability positively affects the probability of success

Market factors

H3a: Market variety positively affects the probability of success

H3b: Retailer willingness and ability positively affect the probability of success

Industry factors

H4a: Information communication technology growth positively affects the probability of success

H4b: E-commerce growth positively affects the probability of success

H4c: Flexible production technology growth positively affects the probability of success

Organizational factors

H5a: The company's level of production technology and flexibility positively affects the probability of success

H5b: The company's level of logistics and distribution system flexibility positively affects the probability of success

H5c: The company's level of customer information dissemination positively affects the probability of success

H5d: First-mover advantage positively affects the probability of success

H5c: The available resources and company's readiness positively affect the probability of success

Figure 2-9. Success factors for achieving mass customization by Broekhuizen/Alsem
(Broekhuizen/Alsem 2002, pp. 319)

4.2   A Comprehensive Framework Encompassing the Main Conditions for Achieving Mass Customization

The presented literature review reveals that the authors do not fully agree with the scope and extent of the needed skills and conditions that must be available in order to successfully achieve mass customization. Although all authors deal with the same issue, the terms used are quite different. Da Silveira et al. (2001) and Broekhuizen/Alsem (2002) speak about "success factors", whereas Zipkin (2001) and Berman (2002) about "key capabilities". However, Mchunu et al. (2003) use the term "key competencies" for achieving mass customization. Kotha (1996) and Pine (1993) respectively use the terms "necessary conditions for success" and "demand and structural factors", whereas Hart (1995) prefers the term "key decision factors".

The contributions reviewed emphasize the basic conditions justifying the shift to mass customization. They generally assume that a supplier with an established production system wants to move onward to mass customization. This is in accordance with the work of Reiss/Beck (1994, pp. 28) who examine the different roads leading to mass customization. The authors argue that there are two principal alternatives. Either the mass manufacturer can shift to mass customization by attempting to preserve cost efficiency or the craft customizer can implement the strategy by maintaining the differentiation advantages. However, the framework to be elaborated should not only emphasize the conditions to be satisfied when moving into mass customization, but also those that are required for its pursuit. In so doing, we guarantee that the framework is more comprehensive and more helpful for managers because it explicitly distinguishes between both cases, namely before and after moving into mass customization.

For the elaboration of the framework, we primarily focus on *what* conditions are necessary for the implementation and pursuit of mass customization. We do not concentrate on *how* these conditions can be fulfilled or influenced. For instance, we argue that technology is just a means that supports manufacturing flexibility or optimal customers' needs elicitation. Technology pertains to the category of enablers which can be defined as "… the means by which change occurs" (Hart 1995, p. 41). That is why, factors that relate to technology are not included in the framework. The framework summarizing the main conditions that we believe to be decisive for the achievement of mass customization is represented by figure 2-10. It distinguishes between market conditions and customizing ability as main dimensions before shifting to mass customization. After implementation, several internal abilities of the mass customizing system have to be maintained and further developed.

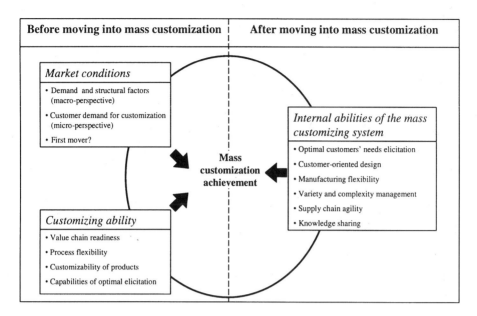

Figure 2-10. Necessary conditions for achieving mass customization

4.2.1 Market Conditions

Before shifting to mass customization, the supplier has to carefully examine whether the market conditions are favorable with respect to (a) demand and structural factors, (b) customer demand for customization and (c) first-mover advantage.

Demand and structural factors

We argue that the market turbulence map of Pine (1993, pp. 66) that consists of the demand and structural factors is a suitable tool. Pine (1993, p. 73) proposes a market turbulence questionnaire to be distributed to key executives and/or knowledgeable managers and professionals across a broad section within the company. On the basis of their responses, the market turbulence map can be drawn up, from which the turbulence level of the company's market environment can be determined. However, the market turbulence map of Pine singly takes into account information that stems from company members and not from customers.

Although Pine (1993, p. 71) empirically validates his market turbulence model by examining the correlation between the market turbulence factors and measures of customization and variety, we think it is also important to evaluate whether the customers themselves are interested in customization.

Customer demand for customization

The success of mass customization should be primarily initiated by customers. Their willingness to have individualized products with eventual premium prices and longer delivery times is a decisive condition that has to be examined before moving into mass customization. It is obvious that if final customers do not have any interest in customization, pursuing mass customization strategies will be superfluous. Zipkin (2001, p. 86) points out that customization may truly add value to customers when they sharply differ in their preferences for certain product attributes. MacCarthy et al. (2002, pp. 76) define the concept of key value attributes for mass customization as the attributes with the greatest perceived value to the customer. The analysis of the key value attributes is a relevant issue in order to determine where the focus of product variety should be. It can be identified as to which product variety is important and value adding from the customers' perspective.

However, a large body of research examining when mass customization would represent a serious opportunity including both suppliers' and customers' perspectives does not exist. "To date, very little scientific work examines consumer behavior and attitudes toward customized products. [...] In particular, we need to know how much consumers care for customized offerings and which customized products or services would be more wanted by consumers" (Guilabert/Donthu 2003, p. 2). In order to contribute in the filling of this research gap, Guilabert/Donthu (2003, pp. 3) use the notion of *Customer Customization Sensitivity* (CCS) which was first introduced by Hart (1995, p. 40). The authors define CCS as the *"customer's susceptibility to preferring customized products/services"* (Guilabert/Donthu 2003, p. 3) and develop a scale consisting of six dimensions which help evaluate whether potential customers will accept customization or not. These dimensions are presented in figure 2-11.

Customer Customization Sensitivity (CCS) Scale
1. In general, customized products/services meet my needs better than standard ones
2. I wish there were more products/services that could be easily customized to my taste
3. I believe there is a need for more customized products/services
4. If the price is similar for standard and customized products/services I would choose customized products/services
5. If I have to wait the latest version of a "---" product service I'd go with the previous version instead
6. If I have a choice, I prefer to have customized products/services

Figure 2-11. Dimensions of Customer Customization Sensitivity
(Source: Guilabert/Donthu 2003, p. 7)

First-mover advantage

The last factor pertaining to market conditions is the first mover advantage. Kotha (1996, p. 447) argues that being a first mover in implementing mass customization will be beneficial for the supplier's image. Even when competitors enter the mass custom segment afterwards, they will find it difficult to prevail, especially when customers consider the first mover as the leader and best supplier of individualized products.

4.2.2   Customizing ability

Customizing ability deals with the evaluation of the readiness of the whole organization before moving into mass customization. Customizing ability basically must be examined with respect to (a) the value chain readiness, (b) process flexibility, (c) customizability of products and (d) capabilities to optimally elicit customers' requirements.

Value chain readiness

The objective of evaluating the value chain readiness before shifting to mass customization is to examine the customizing potential of the entire network including the company, its suppliers, distributors and retailers. It is important to note that companies generally reduce the vertical range of manufacture and focus on core competencies (e.g. Wigand et al. 1997, pp. 190) which are decisive for success in competition. Consequently, a large number of parts required in the end product assembly might be outsourced. Thus, it is conceivable that other partners in the value chain take over the task of producing the customized elements or the customization of elements. For this reason, several partners in the value chain can play a substantial role in achieving good responsiveness to customers' preferences. The result of the evaluation of the value chain readiness is a specification of the activities of the individualization process to be carried out inside the company and those to be outsourced. If strategic suppliers, distributors or retailers do not have the necessary capabilities, a narrow cooperation with the objective to improve the customization capabilities of the entire value chain, is required before moving into mass customization.

For example, Schenk/Seelmann-Eggebert (2003, p. 7) introduce four different logistical approaches of how to realize mass customization strategies across the value chain. They discuss different scenarios where the customizing process can be carried out either by the producer, logistics provider, retailer or customer (Figure 2-12). The authors emphasize that all partners have to be involved at an early stage before introducing any of the four mass customization approaches.

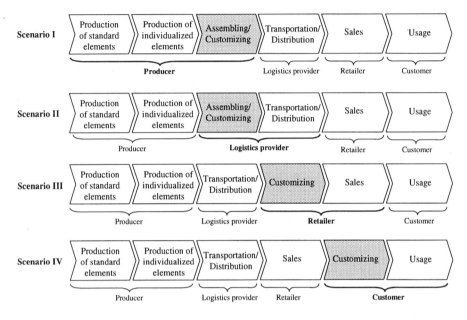

Figure 2-12. Customizing positions on the value chain
(Schenk/Seelmann-Eggebert 2003, pp. 3)

However, Hoogeweegen/Hagdorn-van der Meijden (2003, pp. 2) mention that pre-formed stable business networks of organizations are no longer available by introducing mass customization. Instead, dynamic business networks characterized by high instability emerge because an actual formulated customer order determines which organizations have to be involved in the fulfillment of this particular order. On the basis of the business networking game, a business network consisting of many companies is simulated. As the customers' demand shifts to more product customization, the different organizations (e.g. suppliers, service providers, logistics providers) involved in the network have to adjust their strategies by e.g. acquiring more competencies, specializing their offers or building new relationships with other organizations. The game shows that the organizations do not equally profit from the customization trend. Depending on the pursued strategy, some organizations improve their profit share, whereas others suffer a loss. These results have important managerial implications. They suggest that the shift of the entire value chain to mass customization may be advantageous for some organizations and disadvantageous for others.

Process flexibility

In order to cope with a high diversity of customers' requirements in mass customization a large product variety is needed. This induces a variant-rich production where process flexibility plays a decisive role. Process flexibility can be improved by minimizing setup times. A setup can be defined as "...any changeover activity that is necessary in batch manufacturing to change parts, fixtures, tooling, equipment programming, or instructions from one product, or product variation, to another" (Anderson 1997, p. 46). Before moving into mass customization it is relevant to make sure that the current production system can be easily adapted to use flexible processes and to incorporate computer-aided manufacturing (Berman 2002, p. 59).

Customizability of products

In addition, to make processes flexible some changes on the product design level are necessary. In fact, "[t]he concept of modularity is a basic building block in the manufacturing situations traditionally considered to be flexible" (Duray et al. 2000, p. 610). Moreover, modularity is an approach that enables manufacturers to postpone customers' orders. Modules can be anonymously produced on a mass scale. When a customer order arrives, modules are then combined in a way that will satisfy the specific requirements of the customer. Thus, the product modularity level not only influences manufacturing flexibility but also the customizability of products.

Capabilities of optimal elicitation

Before moving onward to mass customization, the necessary capabilities for optimal customers' needs elicitation should be available. Kotha (1996, p. 447) emphasizes that mistakes and errors in processing customers' orders can be extremely costly in mass customization. Therefore, the company has to implement mechanisms that help customers find the products that fully correspond to their requirements. The information gained from customers is of high relevance since it represents the basis on which the individualized product is manufactured. If a mistake occurs at the elicitation stage, then the product will by no means correspond to customers' expectations.

In order to have access to customer information, it may be required to develop an inter-connected information network with a group of retailers who maintain direct contact with customers. Another alternative is to directly communicate with customers by e.g. implementing a software tool over the World Wide Web which provides customers with the possibility to change and visualize product variations. The supplier should consider many alternatives and evaluate them with respect to the available capabilities and potential for the successful achievement of mass customization.

4.2.3   Internal Abilities of the Mass Customizing System

The internal abilities of the mass customizing system are the skills and capabilities that must be maintained and further developed when pursuing mass customization. These are: optimal customers' needs elicitation, customer-oriented design, manufacturing flexibility, variety and complexity management, supply chain agility and knowledge sharing. It is important to note that the mass customizing system includes not only the firm who is liable for the individualized product but also its partners in the value chain that actively participate in the customization process.

Optimal customers' needs elicitation

In order to be able to offer a superior value to customers, the mass customizer has to continuously develop its capabilities with regard to customers' needs elicitation. There are four kinds of elicited information in mass customization: identification, such as name and address; customers' selections from product alternatives; physical measurements; and reactions to prototypes (Zipkin 2001, p. 83).

The information related to identification and customers' selections can be easily gained by implementing software systems over the Internet. Nowadays, these software systems that are sometimes called configurators can be cheaply and quickly constructed. They guide customers through an array of choices with the objective to lead them to the suitable product variation. In order to improve learning about customers and optimize the elicitation process, the integration of configurators with Customer-Relationship Management (CRM) systems is very promising. CRM systems aggregate customer related data and derive useful information such as individual desires and behaviors. This information can be used to help customers find in a fast paced manner the product that fully meets their requirements (Blecker et al. 2004a, pp. 7).

In many cases, the customization process requires information about physical measurements. For instance, body measurements are necessary when the product is a custom-made pair of trousers. This information can be elicited either manually (manual measurement of waist and length by oneself or at a retailer store) or automatically (automated body measurement through an optical body scanner).

However, the use of prototypes to elicit customer information is not yet a common method and requires complex software systems. Some application fields stem from the building industry where customers are invited to build their houses with a click of the mouse (Zipkin 2001, p. 83).

Customer-oriented design

A fundamental internal ability that has to be available when pursuing mass customization is an efficient customer-oriented design. Before starting to listen to the "voice of the customer", one must determine which "voices" to listen to, which means to decide which market segments might be attractive for customization. Furthermore, "[i]n contrast to the development of a discrete product, developing products for Mass Customization focuses on product *families/platforms* or many product families and their *evolution over time*" (Anderson 1997, p. 202).

Thus, in mass customization, customer requirements have to be mapped to a family of products, instead of a single product such as in mass production environments. The mass customizer must determine the product attributes for which customer requirements are considerably different. Therefore, a product family generally consists of both standardized elements (product platform) and variable elements to ensure product variation. It is noteworthy that product design to a great extent determines the costs in manufacturing. Each product variation is associated with additional costs and therefore too much variance can weaken the entire product program.

Manufacturing flexibility

Slack (1983, pp. 4) argues that manufacturing flexibility has three main dimensions: (a) the range of possible configurations a manufacturing system can adopt, (b) the cost of migrating from one configuration to another, and (c) the time needed to make a transition. According to this definition, a mass customization manufacturing system should endue a wide range of possible configurations that makes the production of a large product variety possible. Moreover, both the costs and time necessary to change from one configuration to another (i.e. from one product variation to another) should be kept low, which means that the setup costs and times have to be minimized. In this context, Pine (1993, p. 50) graphically shows the importance of the reduction of set up costs in order to move the Economic Order Quantity (EOQ) towards a lot size of one, which allows the company to efficiently handle more product variety.

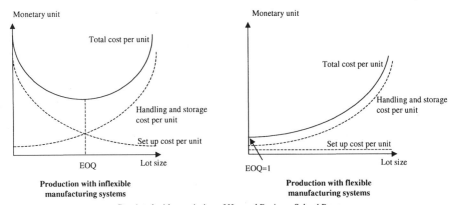

Figure 2-13. The effect of reducing setup costs on economic order quantities
(Adapted from: Pine 1993, p. 51)

When producing with inflexible manufacturing systems, EOQ is the point at which the increasing handling and storage costs begin to outweigh the decreasing setup cost, which yields the lowest total cost per unit (left side of figure 2-13). For flexible manufacturing systems with low setup costs, the lowest total cost corresponds to a batch size of one (right side of figure 2-13). Nevertheless, the achievement of "a lot size of one" is very demanding and is by no means a simple task even if modern manufacturing technology is available. Therefore, during the pursuit of mass customization, manufacturing flexibility has to be continuously improved by minimizing both setup times and costs.

Variety and complexity management

This condition that is necessary for the achievement of mass customization is not mentioned in the literature reviewed in section 4.1. We demonstrate in the next chapter that variety and complexity management are indispensable pillars for the pursuit of mass customization. Variety management embraces methods that control product variety proliferation. The main concern is to cope with the complexity stemming from the product (e.g. number of parts, components, and variants) as well as the complexity acting on the product (e.g. market diversity, production flows) by means of suitable instruments. Complexity management not only deals with the proliferation of product variety, but also with the proliferation of processes and resources in companies. Through a conscious increasing or decreasing of complexity, the target is to cope with variety in all value adding chain activities, so that a maximal contribution to customer perceived value and a high efficiency are simultaneously attained (Schuh/Schwenk 2001, pp. 34-35).

Supply chain agility

Fisher (1997, pp. 106) develops a taxonomy based on two main dimensions for matching supply chains with products. The first dimension distinguishes between functional and innovative products, whereas the second dimension makes the difference between efficient and responsive supply chains.

Functional products satisfy basic needs and do not change much over time. They have a stable, predictable demand and long life cycles. However, innovative products have short life cycles. Their very newness makes demand for them unpredictable. In addition, innovative products are characterized by a great variety which further increases unpredictability.

Efficient supply chains basically focus on the reduction of physical costs which include the costs of production, transportation, and inventory storage. In opposition, responsive supply chains concentrate on minimizing market mediation costs that arise when supply exceeds demand or when supply falls short of demand. The taxonomy presented by figure 2-14 suggests that a match between supply chains and products occurs when functional products are manufactured and delivered within the scope of efficient supply chains, whereas innovative products are made within a responsive supply chain. Otherwise, there is a mismatch.

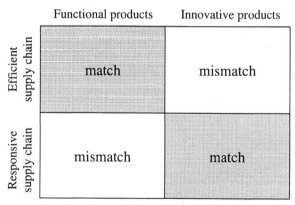

Figure 2-14. Matching supply chains with products

(Source: Fisher 1997, p. 109)

All properties that Fisher has used to define innovative products also hold for mass customized products. Therefore, good matching assumes that mass customized products are produced within responsive supply chains. In addition, Goldman et al. (1995, p. 330) define agility as the rapid system responsiveness to unforeseen customer requirements. From Fisher's taxonomy and the agility definition of Goldman et al., it can be deduced that mass customized products require supply chain agility. Furthermore, Naylor et al. (1999, p. 112) argue that supply chain agility gains importance in manufacturing environments characterized by a high demand for product variety and a high demand for variability in production.

To achieve supply chain agility, the mass customizer has to manage uncertainty that is triggered by unforeseen customer requirements. Three coordinated strategies can be employed, namely by (a) reducing uncertainty (e.g. finding sources of new data), (b) avoiding uncertainty (e.g. cutting lead times and increasing supply chain's flexibility), and (c) hedging against the remaining residual uncertainty (e.g. with excess capacity) (Fisher 1997, p. 14).

Knowledge sharing

Mass customization is a dynamic strategy that highly depends on the ability of translating new customer demands into new products (Da Silveira et al. 2001, p. 4). Knowledge about customers' preferences and desires that is obtained at retail stores or through a direct interaction with customers has to be rapidly and efficiently transferred across the value chain. Moreover, the achievement of mass customization considerably relies on the ability to create and harness a worker's knowledge at the point of production. Compared to mass products, the manufacturing of mass customized products requires higher skills and qualifications of employees (Kotha 1996, p. 448).

5.   THE CHALLENGING APPROACH OF PRODUCT CUSTOMIZATION: MASS CUSTOMIZATION

In the technical literature, there is no consistent and uniform definition for mass customization. This may be due to the fact that mass customization is a paradigm which has quite recently gained more attention from academia and practitioners. Research on mass customization is not extensive and further work is needed to solve many research questions. The concept emerged in the late 1980s and the term "mass customization" was coined for the first time by Davis (1987) in his book "future perfect". The popularity of the concept has dramatically increased among managers and academics after the publication of Pine's book "Mass Customization: The New Frontier in Business Competition" by 1993.

Another reason why mass customization lacks a generally valid definition is the multidisciplinarity of the approach. The work on mass customization involves many researchers from different fields such as business administration, artificial intelligence, mechanical engineering, industrial engineering, and psychology, just to name a few. Thereby, each field has developed specific language in order to deal with research issues in mass customization.

In order to define mass customization, Pine (1993, p. 44) notes that "...practitioners of Mass Customization share the goal of developing, producing, marketing, and delivering affordable goods and services with enough variety and customization *that nearly everyone finds exactly what they want.*" Pine basically focuses on the value chain that should have the ability to provide customers with choice and individualization possibilities, so that nearly every customer finds the product corresponding to his or her needs.

Piller (2000, p. 206) defines mass customization as the production of goods and services for a relatively large market, in which the needs of each individual customer can be met at a cost level that is comparable to that of mass producers. The information that is gained from customers during the customization process serves as a means to build long-lasting relationships with each customer. The definition set forth by Piller consists of two main parts. Whereas the first part deals with the production of individualized goods and services with near mass production efficiency, the second part rather emphasizes on the learning relationship between customers and mass customizer. The involvement of customers is a necessary requirement for product customization and the information gathered can be used in order to more efficiently meet customer preferences and desires. For instance, Broekhuizen/Alsem (2002, p. 311) do not agree with Piller that learning relationships are important to efficiently pursue mass customization. They point out that when the time gap between repurchase is considerable, the information gained from customers may become irrelevant. However, we believe that the knowledge acquired from customers is always of great relevance. The main concern should be attributed to the implementation of mechanisms and procedures that enable the evaluation of this information. At a specific point in time, only useful and filtered information has to be exploited in order to better fulfill a customer's requirements, whereas obsolete information has to be discarded.

In opposition to Pine (1993) and Piller (2000) who each give a single definition for mass customization, Hart (1995, p. 36) argues that two distinct definitions would be necessary for mass customization. The first one is called the visionary or Platonic definition of mass customization. It is " ... the ability to provide your customers with anything they want profitably, any time they want it, anywhere they want it, any way they want it" (Hart 1995, p. 36). The second definition is the practical one. It is "the use of flexible processes and organizational structures to produce varied and often individually customized products and services at the low cost of a standardized, mass production system (Hart 1995, p. 36). Whereas the first definition rather represents a goal or an ideal that can be rarely achieved, the second one is pragmatic and can be understood as customization within a predetermined "envelope of variety" (Hart 1995, p. 37). The role of the visionary definition is to continuously incite suppliers to intensify and reinforce their customer orientation. We argue that Hart has defined mass customization in an intelligent way. On the one hand, he emphasizes the strong customer orientation that mass customization has to strive for. On the other hand, he does not neglect the efficiency constraint that also has to be satisfied when pursuing mass customization.

Thus, what is common among all of the definitions discussed above is that mass customization has to include two main perspectives, namely customer orientation and costs' efficiency. This is also in accordance with the definition set forth by Duray et al. (2000) who argue that customer involvement in the production cycle as well as modularity to achieve efficiency are the most important dimensions to classify mass customizers. We also argue that mass customization has to be built upon these two fundamental pillars. For this reason, we totally reject the claim of Da Silveira et al. (2001, p. 4) who state that modularity is not the fundamental characteristic of mass customization and that true mass customized products are individually made.

We define mass customization as the process of the fulfillment of the needs of each individual customer pertaining to a relatively large part of a market the supplier focuses on, with prices that do not exceed 10-15% of the prices of comparable standard products.

Furthermore, mass customization has to provide each customer with the required means enabling him or her to easily and comfortably interact with the product assortment in order to find the product corresponding to his or her needs. The information that arises during the customization process serves on the one hand to build a long-lasting relationship with the customer and on the other hand to continuously optimize operations and manufacturing-related tasks in order to improve costs' efficiency.

6.   SUMMARY

In this chapter, we define product customization and we show its contribution to create a competitive edge over competitors in dynamic business environments. To customize products, two main alternatives are identified. The first alternative is to use craft customization, whereas the second one is to implement mass customization. Craft customization is achieved by methods and concepts from the project management field. It reposes on a job shop manufacturing system characterized by extremely high flexibility. However, mass customization is an emerging paradigm that is challenging for manufacturing enterprises. In comparison to craft customization, mass customization assumes a more restricted level of flexibility. It aims at individualizing and delivering products by means of processes that nearly have mass production efficiency. Whereas craft customization is a well-established concept that has been applied for many decades, mass customization is still a new paradigm that requires further research. Therefore, within the scope of this book, we argue to uniquely handle mass customization.

Mass customization can be considered to be a strategy in itself. However, the literature research has shown that there are many strategies or configurations for the pursuit of mass customization. In the contributions reviewed, the authors classify mass customization concepts differently. The main conclusion to be drawn is that mass customization configurations can be represented on a continuum with adaptable standard products at one extreme point and products incorporating a very high level of customization at the other extreme point. Different classification approaches as discussed in the literature are criticized according to specific criteria but we do not provide a categorization model. We are convinced that each model in the literature reflects a particular definition of mass customization and therefore a specific perspective when dealing with mass customization. For example, some researchers would consider the manufacturing of adaptable standard products as a form of mass customization, while other researchers would not do so. For the purpose of this book, the classification model of Duray et al. (2000) will be adopted for the classification of mass customization configurations.

The second literature review examines the necessary conditions for the achievement of mass customization. On the basis of the performed literature research, a framework has been established that distinguishes between the necessary conditions for shifting to mass customization and the necessary conditions to be maintained and further developed while pursuing mass customization.

Finally, we elaborate a working definition for mass customization on which we will base this book. We define mass customization as the process of fulfillment of the needs of each individual customer pertaining to a relatively large part of a market that the supplier focuses on, with prices that do not exceed 10-15% of the prices of comparable standard products. This definition basically concentrates on the customer's perspective. However, it does not neglect the efficiency perspective. In contrast to other definitions, it also explicitly mentions how much "mass" should be put into mass customization in order to be able to speak about mass customization. When the prices of customized products exceed 10-15% of the prices of comparable standard products, then it is no longer legitimate to speak about mass customization. Instead, it will be more suitable to use another term such as job shop production.

Chapter 3

MASS CUSTOMIZATION AND COMPLEXITY

The strategic benefits of mass customization have been widely discussed in the theory of business management. However, large deficits deal with the practical application (Piller/Reichwald 2002, p. 1) because moving into and practicing mass customization represents a very difficult task (Hart 1994, p. 36), especially for suppliers who are accustomed to practicing their business by applying the rules of traditional management concepts. In this context, a relevant reason ascribed to the failure of some mass customization projects is the increasing complexity problem.

Research that examines complexity in the specific case of mass customization is still missing. Up to now, it is very common that one extrapolates the findings of studies on variety and complexity that are achieved in batch or even mass production in order to point out the effects of complexity in mass customization. This point of view is not correct because mass customization has some particularities that should be taken into account when dealing with the complexity issue.

Mass customization induces a high complexity level because of various customer requirements and a steadily changing environment. However, it has some potential to reduce complexity. In this chapter, the interdependencies between mass customization and complexity will be discussed in order to demonstrate that mass customization is not just an oxymoron linking two opposite production concepts, namely customization and mass production, but also a business strategy that can actually lead to success. After presenting a literature review on complexity, we will deal with both of the complexity faces of mass customization. On the one hand, mass customization increases production program, manufacturing and configuration complexities. On the other hand, mass customization can contribute to a reduction in complexity at the levels of the order taking process, product and inventories. Then, the main results attained through the analysis are integrated in a comprehensive framework that shows the complexity increasing and decreasing aspects when introducing and pursuing mass customization.

1.   COMPLEXITY: A LITERATURE REVIEW

Up to now, the term complexity has no satisfactory and generally admitted definition. It is basically discussed in connection with the system theory and is referred to as a system attribute. A system consists of elements or parts (objects, systems of lower order, subsystems) and the existing relationships between them. It is also argued that a system should perform a specific function and have to be well distinguished from its environment without confusion. The complexity of a system is defined with respect to three complexity variables, namely number, dissimilitude and states' variety of the system elements and relationships. These variables enable one to make the distinction between structural and dynamic complexity. Whereas structural complexity describes the system structure at a defined point in time, dynamic complexity represents the change of system configuration in the course of time. For example, by considering the solution space of the mass customizer that consists of all theoretically possible product variations, the product configurations that can be manufactured at a point in time determine the structural system complexity. However, the dynamic complexity basically depends on the frequency and magnitude of changes in the solution space when new product variants are introduced or eliminated.

On the basis of the structural and dynamic complexities, Ulrich/Probst (1995, p. 61) have determined a taxonomy for system complexity. When both complexities are low, then the system is simple. In the case of a high (low) structural complexity and low (high) dynamic complexity, the system is considered to be complicated (relatively complex). When both complexities are high, then the system is said to be extremely complex (Figure 3-1).

Saeed/Young (1998) define complexity in companies as the "...systemic effect that numerous products, customers, markets, processes, parts, and organizational entities have on activities, overhead structures, and information flows". The main problem triggered by too much complexity is the hidden costs. The costs of complexity are generally not visible and can negatively affect the competitive advantage of the enterprise. Mass customization triggers high complexity because of the variety of products, markets ("Markets of one"), processes, customers, etc. However, the mass customizing system cannot be a simple one owing to the complexity of its environment. This is in accordance with Ashby's law of requisite variety in cybernetics, which says that "variety can destroy variety" (Ashby 1957, p. 207) and can be also extended to "complexity can destroy complexity". This is basically due to the fact that a corresponding complexity in the system must be available in order to cope with the complexity of the environment. But the problem remains to determine how much complexity is optimal. Saeed/Young (1998) propose to distinguish between the complexity the customer rewards and the complexity the market is not willing to pay. Frizelle/Efstathiou (2002) also make such a distinction and call the former "good complexity" and the latter "bad complexity".

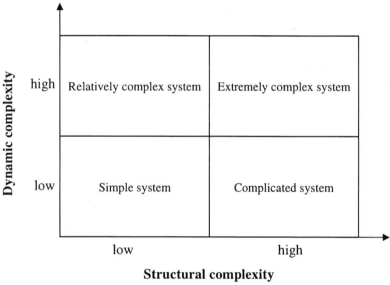

Copyright by Haupt Berne.

Figure 3-1. Complexity of systems
(Adapted from: Ulrich/Probst 1995, p. 61)

In order to cope with complexity, Wildemann (2000, p. 7) identifies three measures to be taken, which are: complexity reduction, complexity prevention and complexity control. Complexity reduction aims at simplifying structures for the short term by e.g. eliminating unprofitable product variants or reducing the customer system elements. Complexity prevention is targeted on e.g. developing methods capable of assessing complexity, for instance costs of variety. Complexity control deals with the rest of complexity that cannot be reduced because of environmental complexity such as the diversity of market requirements. To manage complexity, McKinsey prefers to distinguish between instruments for the reduction of program, product and process complexities (Maroni 2001, p. 51). Other authors (e.g. Reiss 1992, p. 40; Hoege 1995, pp. 81) differentiate between complexity decreasing and complexity increasing measures.

Bliss (2000, pp. 197) has developed an integrated four phase concept for the management of complexity, which is based on a theoretical system analysis. The first step is to eliminate the autonomous enterprise complexity. This means to cut internal complexity inside the enterprise that has no correspondence in the environment and subsequently represents a congestion of the Ashby's law of requisite variety. The objective of the second step is to reduce the correlation between internal and external complexities. The basic target is to make internal complexity less vulnerable to environmental changes. The third step deals with a conscious reduction of the perceived market complexity by simplifying e.g. the production program. The fourth step is targeted towards the complexity control by e.g. modularizing the manufacturing process. The relevance of the work of Bliss (2000) is the determination of a sequence according to which complexity management concepts should be applied (Figure 3-2).

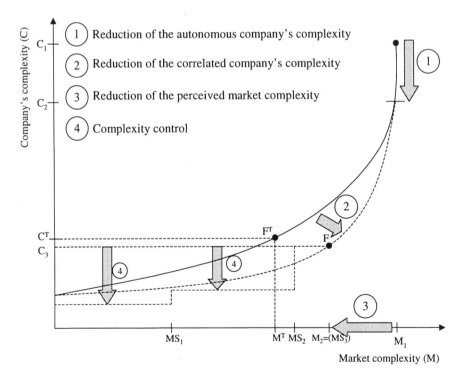

Figure 3-2. Approach of Bliss for the management of complexity

(Source: Bliss 2000, p. 206)

In the technical literature, there are several approaches to manage complexity. These are in some cases contradictory, which emphasizes the strong subjectivity when addressing the complexity problems. But up to now there is not much work that relates to the development of measures or metrics for an objective assessment of complexity. This may be due to the failure and lack of adaptability of many complexity measurements that are suggested in the complexity theory. However, Frizelle/Woodcock (1995, pp. 26) devise an entropic measurement for evaluating complexity in manufacturing. Entropy is well known in thermodynamics and information theory. It provides a measure of the amount of information associated with the occurrence of given states. This measurement, successfully applied in practical manufacturing cases, suggests that complexity reduction can be achieved when there are fewer processes, states and variations of states.

It is noteworthy that the main focus in this chapter is not to develop a concept for complexity management in mass customization or measurements for complexity assessment. This will be elaborated in more detail in chapter nine of this book. The main goal is to discuss and analyze the interdependencies between complexity and mass customization. Some clarifications will be put not only on the causes that trigger complexity in mass customization, but also on the potential of mass customization to reduce complexity.

2.   A SYSTEM VIEW FOR MASS CUSTOMIZING ENTERPRISES

As aforementioned, complexity is always discussed in connection with the system theory. Therefore, in order to analyze complexity in mass customizing enterprises, it is important to delimit at first the system to be focused on. We will now consider the system that contains three subsystems, namely the product configuration subsystem, the manufacturing subsystem and the product arrangement subsystem. We are convinced that this system is the most complex that must be optimized in a mass customizing enterprise (Figure 3-3).

The product configuration system is a software tool for the elicitation of customers' requirements. A distinction has to be made between front end and back end systems. Whereas the front end system is the user interface, the back end system contains the product logic that is the model according to which the product and its variations are represented.

The manufacturing system is required to produce individualized goods with near mass production efficiency. In the technical literature related to mass customization, it is common that one makes the distinction between the mass production and customization components inside the manufacturing system. In addition, it is assumed that a single decoupling point separates between both manufacturing components. Although in practice it is uncommon to encounter manufacturing systems with one *single* point of customization, the decoupling point has a considerable amount of theoretical relevance. It provides an efficient measurement to assess the degree of customization that the supplier is able to provide. It also reflects the level of customer involvement in the production process and enables one to make a comparison between different mass customizers with respect to customer integration. Thus, from a systemic theoretical point of view, the distinction between both the mass production and customization systems is legitimate.

To be able to configure and manufacture mass customized products, product structure is necessary. The product structure is represented inside the product arrangement system which according to Nilles (2002, p. 56), consists of the functional and building-oriented systems. The elements of the functional system are the product functions. This system is essential because customers naturally express their needs in terms of functional requirements. The building-oriented system has technical relevance and enables the mapping of functional requirements into a product-oriented description.

It is important to point out that there are mutual relationships between the different specified systems. The representation of the front end system considerably depends on the functional oriented system. The front end takes over the task of communicating the existing product functions to customers in a user-friendly manner. This is relevant because customers generally ask for product solutions to satisfy a particular need that can be captured in terms of functions but not in terms of product components.

In the back end system, the components and product constraints ensuring consistent product variations are modeled. The elaboration of product models considerably depends on the product architecture (building oriented system) conceived by engineers at the design stage.

In the building oriented system, the platforms as well as the modules and components that are necessary for production are specified. This determines to a great extent the machine processing and routings necessary for manufacturing and assembly. Thus, the influence of the building-oriented system on the manufacturing system is obvious. Another important issue is that the capabilities of the manufacturing system should be taken into account during the specification of the product architecture which is modeled in the building oriented system. For example, Design For Manufacturing (DFM) suggests that manufacturing capabilities must be considered at the early stages of the design process. From this point of view, it can be stated that the manufacturing system may involve some constraints or restrictions during the definition of the building-oriented system. In addition, it is conceivable that the configuration system interacts with the planning component of the manufacturing system, in order to retrieve data that is relevant for the determination of e.g. delivery times. The mutual interactions between the configuration and manufacturing systems is what is being debated here.

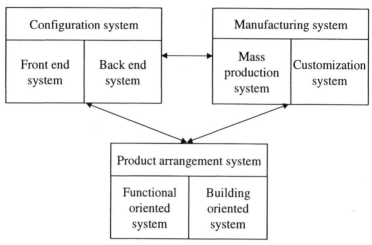

Figure 3-3. The system to be optimized in mass customization

(Source: Blecker et al. 2004d, p. 893)

3.   INCREASING COMPLEXITY DUE TO MASS CUSTOMIZATION

3.1   Mass Customization Triggers High Production Program Complexity

The production program consists of all products that are manufactured in the enterprise. In this context, one should make the distinction from the product program which contains not only the manufactured products (i.e. the production program), but also the end products to be sold without being processed (goods for resale). In mass customization, the involved production program is generally characterized by a very high variety. For example, Cmax.com, a mass customizer of sport shoes offered approximately $3*10^{21}$ variants over the Internet. The entire surface of the earth would scarcely suffice for exhibiting all the possible variants of shoes (Piller et al. 2003, p. 6). Additionally, many examples of companies from the automotive industry show that the number of product variants has considerably increased in the last years (Piller 2000, pp. 175).

To illustrate the rapid proliferation of product variety, Rosenberg (1996, pp. 2120) demonstrates this with an example from the automotive industry as to how the total number of product variants that are manufactured on the basis of 9 must-modules and 14 can-modules can go into the billions (Figure 3-4). Must-modules are indispensable for ensuring the basic product functionalities, whereas can-modules are optional. An engine is a must-module because it contributes to the basic functionality of mobility. However, an air-conditioner is a can-module whose absence does not disturb the mobility function.

Must variants (MV)				Can variants (CV)		
m	X_m			k	Y_k	
1	7	Engines		1	1	Front spoiler
2	3	Gear boxes		2	1	Rear spoiler
3	2	Brake systems		3	1	Fog lamp
4	2	Car bodies		4	1	Tachometer
5	2	Chassis		5	1	Multi-functional display
6	15	Exterior colors		6	3	Radios
7	8	Seat covers		7	2	Right exterior mirrors
8	2	Vitrifications		8	2	Sunroofs
9	2	Window lifts		9	1	central locking system
				10	1	Adornment strips
				11	2	Antennae
				12	1	Air conditioner
				13	1	Seat heater
				14	1	Airbag

$$MV = \prod_{m=1}^{9} x_m = 80.640$$

$$CV = \prod_{k=1}^{14} (y_k + 1) = 110.592$$

$$PV = \prod_{m=1}^{9} x_m \prod_{k=1}^{14} (y_k + 1) = 8.918.138.880 \approx 8,9 \text{ Billions}$$

Figure 3-4. Increasing complexity as a consequence of variant-rich production
(Source: Rosenberg 1996, p. 2120)

An empirical study of Wildemann (1995a, p. 14) has shown that with the doubling of the number of product variants in the production program, the unit costs would increase about 20-35% for firms with traditional manufacturing systems. For segmented and flexible automated plants, the unit costs would increase about 10-15%. Wildemann concluded that an increase of product variety is associated with an inverted learning curve (Figure 3-5). This a priori implies that mass customization hardly leads to ensure a competitive advantage because of high variety which is driven by strong product differentiation. This problem is also complicated by the fact that customers do not accept high prices even when receiving individualized goods. Moreover, many studies have confirmed that manufacturing enterprises with a narrow production program are more successful than those with broader range of product variety.

Variety-induced complexity triggers higher costs which arise in the form of overheads (e.g. Anderson 1997, pp. 45; Rosenberg 2002, pp. 225). In addition, even by applying some modern cost calculation concepts such as activity based costing, it is generally very difficult to adequately and fairly allocate these indirect costs to the corresponding product variants.

Furthermore, the development of modular product architectures is very time-consuming and cost-intensive. A leading problem is to specify the module and platform interfaces. The number and the variety of interactions between modules in the building-oriented system of the product arrangement system, increase the production program complexity.

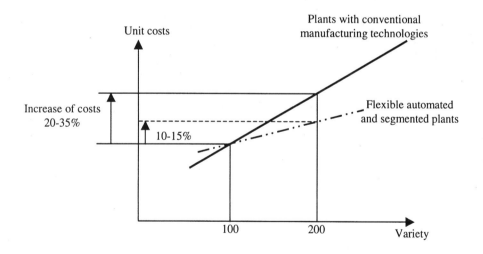

Figure 3-5. Inverted learning curve with variety doubling
(Source: Wildemann 1995a, p. 14)

3.2   Mass Customization Triggers High Manufacturing Complexity

Mass customization strategies (fabricators, involvers, modularizers, assemblers) involve a high product variety. This increasing product variety triggers a principal challenge in manufacturing, which is to efficiently plan and control production. In such an environment, PPC (Production Planning and Control) systems such as MRPII (Manufacturing Resource Planning) or ERP (Enterprise Resource Planning) that are originally designed to support manufacturing in operations with a limited number of product variants, are not efficient (Rautenstrauch 1997, pp. 401; Tseng/Jiao 2001, pp. 13). The primary encountered problems basically lie in the necessity to specify all of the possible product variants in the system.

Furthermore, mass customization requires flexible manufacturing systems on the shop floor. With such systems, it is possible to manufacture high product variety in little batch sizes at relatively low costs. Although flexible manufacturing systems with flexible alternative machines, operation sequences and routings lead to potential improvements in the manufacturing system performance, they involve significant increases in the size of the production planning problem. Byrne/Chutima (1997, p. 110) point out that an added degree of freedom due to manufacturing flexibility enormously increases the complexity of the structure of the scheduling function. Thus, flexible manufacturing systems for mass customization do not entirely solve the problem caused by variety because of high planning and scheduling complexity.

3.3   Mass Customization Triggers High Configuration Complexity

In opposition to the mass production system, where the manufacturer tells customers what they buy, mass customization assumes that consumers tell the manufacturer what to produce (Tseng/Jiao 2001, p. 16). Furthermore, in the technical literature the customer is considered to be a "prosumer", "co-producer" as well as "co-designer". This points out that in mass customization, the enterprise system should not only involve the internal value adding system as it is common from a traditional view, but also the customer system (Figure 3-6). The traditional view would be satisfactory in business environments where customers do not undertake an active role in the value chain. The mass customization view considering the internal value adding and the customer systems is of high relevance when dealing with the complexity issue.

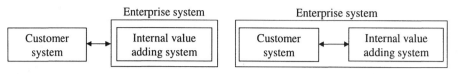

Figure 3-6. The enterprise system from a traditional view and
a mass customization optimal view

(Source: Blecker et al. 2004d, p. 896)

In mass customization, it is common that manufacturers use configuration systems to enable customers to express their needs. von Hippel (2001) even speaks of innovation toolkits which provide customers with the possibility to innovate by way of self-designing products. The customer system has to be strongly considered when dealing with the configuration process because when product variety is high and reaches an astronomical scale, customers have difficulties in making a decision between product variants. They generally feel lost in huge product assortments and are overwhelmed by the configuration task. This aspect is called configuration complexity. Huffman/Kahn (1998, pp. 501) compared the attribute-based and alternative-based presentations of product variants. They find that customers can better discover their preferences thanks to an attribute-based presentation. This means that the configuration complexity depends on the features of the software tool used for supporting customers. Piller et al. (2003, pp. 6) point out that because of large variety, customers are overloaded with information, which can result in configuration processes that take a long time. Moreover, increasing uncertainty in the decision-making process may lead customers to an unwanted behavior, that is, to abort configuration and go away.

4. DECREASING COMPLEXITY DUE TO MASS CUSTOMIZATION

Although mass customization involves increasing complexity at many levels, it has some potential to reduce complexity within the enterprise system. Mass customization involves a very specific business environment in which customers no longer have a passive role in the value chain, but they are able to provide valuable and direct input. The implementation of product configuration systems enables the reduction of the order taking process complexity. Furthermore, it is not necessary to carry final products' inventory because products are not manufactured or assembled until the customer's order arrives. In addition, product complexity is reduced by standardization and modularization.

4.1   Mass Customization Reduces Order Taking Process Complexity

The implementation of product configuration systems over the Internet provides customers with the possibility to configure their products according to their requirements and to send their orders per mouse-click to the manufacturer who can begin with production. Customers interact with the front end of the configuration system in order to express their needs. They can also visualize their choices and change them according to their specific requirements. The back end system of the configurator prohibits inconsistencies between parts or modules, so that only producible variants are allowed to be ordered. Forza/Salvador (2002, p. 95) point out that errors during order acquisition can be considerably reduced with the introduction of product configuration systems. Product variant prices as well as delivery point in times can also be automatically determined. Furthermore, product configuration systems can provide sales personnel or retailers with valuable support during their interaction with customers.

The integration of the configuration system with e.g. the product data management (PDM) system and the PPC system provides additional advantages. Product documentation with respect to involved parts or modules and routings can be automatically and efficiently generated (Blecker et al. 2003, pp. 21). Product configuration systems generally do not attribute a different part number to each product variant. This would induce an explosion of data because of the possible variety of customers' orders. Therefore, configurators use a generic product structure that enables one to efficiently represent product data by avoiding redundancies (Tseng/Jiao 2001, p. 12).

4.2   Mass Customization Reduces Product Complexity

As aforementioned, modular product design is a very relevant issue in mass customization. Although modular architectures may induce some complexity during the design task, especially with respect to the specification of interfaces, they are decisive in making mass customization function efficiently. Piller/Ihl (2002, pp. 13) ascribes the development of mass customization to the advances realized in the field of modular designs. However, it is noteworthy that product modularity does not necessarily imply that the supplier pursues mass customization. For example, some automobile manufacturers produce cars around modular architectures but still receive orders from retailers who do not imperatively involve specific customers' requirements. From this point of view, modularity is just an enabler for mass customization.

Conversely, in order to put the "mass" in mass customization, a mass production system has to be involved. For this reason, the product has to be designed in such a way that mass production is possible, which can be optimally realized by developing modular architectures. That is why, mass customization obligatorily implies modular designs (Figure 3-7), which means a reduction of product complexity. So, modules can be manufactured independently from customer orders within a mass production system. After customers specify their requirements, modules are assembled together into product variants within the scope of the customization process. The concept behind this organizational approach, which was made possible by modular designs, is called postponement (e.g. van Hoek 2001, pp. 161) which means to delay some activities in manufacturing until the customer's orders are received.

Modular architectures reduce product complexity by maintaining large end product variety. By opting for an integral architecture instead of a modular design, the potential to manufacture billions of product variants as it is common in mass customization would mean to design billions of different variants. Ulrich (1995, p. 427) points out that a fully integral design may require changes to every component in order to be able to effect change in any single functional element of the product. In opposition, modular structures enable one to manufacture a high range of variety on the basis of only a few modules. Modular architectures increase the commonality level between products, which improves manufacturing performance. In addition, mass customization makes designers strive for increasing components commonality in- and between different modules.

Due to decoupled interfaces, a modular product architecture enables one to change certain modules for e.g. an upgrade without having to introduce changes in surrounding modules. Moreover, in contrast to an integral architecture, a modular design does not necessarily assume high flexible production equipment in order to efficiently achieve a high range of product variety. Another positive aspect of product modularity is that it enables a better exploitation of supplier capabilities (Ulrich 1995, p. 430).

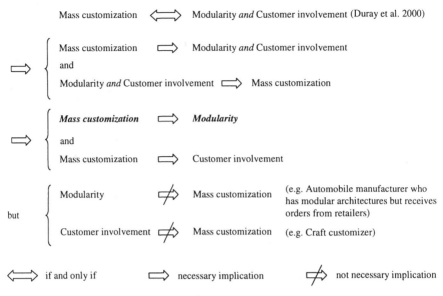

Figure 3-7. Mass customization implies modularity

4.3   Mass Customization Reduces Inventory's Complexity

Classical mass production is based on requirements' forecasting, which generally means high inventories because of "product-push". However, mass customization reposes on a "customer-pull" strategy with the main advantage that production does not begin until the customer's order arrives. The result is that end product inventories are no longer required and the corresponding inventory costs are avoided. Industrial sectors that suffer from high customer demand fluctuations can reduce sales planning complexity by implementing mass customization. This will improve sales planning reliability. However, the make-to-order concept in mass customization assumes that customers do not immediately receive their configured products. Nevertheless, customers accept a certain delay between order and delivery because they value the additional benefit provided by customized products.

The implementation of flexible manufacturing systems in mass customization decreases complexity at the work-in-process inventory level. It is true that these modern manufacturing systems induce high scheduling problems, but the basic advantage consists in their ability to considerably reduce setup times and manufacturing lead times. Furthermore, as aforementioned, mass customization encourages the standardization of components by increasing the commonality level in- and between product variants. The result is a decreasing complexity of the work-in-process inventory. Thereby, it is not only the inventory volumes, but also the number of part and component types in the inventory that are decreased.

5. INTERDEPENDENCIES BETWEEN MASS CUSTOMIZATION AND COMPLEXITY

Mass customization is a strategy that not only increases complexity in the enterprise system but also with some potential to decrease complexity. On the one hand, mass customization can yield increasing complexity at the configuration, planning and scheduling, and production program levels. We call these aspects complexity drivers. On the other hand, mass customization strongly contributes to the reduction of complexity of the order taking process, inventory and product. These aspects are called complexity breakers in mass customization.

Before moving into mass customization, a manufacturer has to outweigh the effects of the complexity drivers and breakers. Complexity drivers may induce additional hidden costs that arise in the form of overheads. Complexity breakers will contribute towards a decrease in complexity in the long term. However, the complexity breakers can involve single investment costs when e.g. implementing an online configuration system.

The complexity drivers and breakers in mass customization are integrated in a comprehensive framework (Figure 3-8). The right side of the framework depicts increasing complexity due to mass customization, whereas the left side illustrates decreasing complexity due to mass customization. It is noteworthy that all complexity aspects (configuration complexity, planning and scheduling complexity, and production program complexity) can all be ascribed to a single cause, namely the proliferation of variety involved by way of mass customization. Thus, in order to optimally cope with complexity in mass customization, a relevant condition is to master and control the proliferation of variety. However, the decreasing complexity is essentially ascribed to the use of configuration systems, customer pull production and set up reduction, and standardization and increasing commonality through product modularity.

It should not be believed that the complexity drivers cannot be influenced. Some measures have to be undertaken in order to reduce the magnitude of their effects. The main concern is to benefit from the inherent complexity decreasing aspects of mass customization (complexity breakers) and to optimally master and control the complexity drivers.

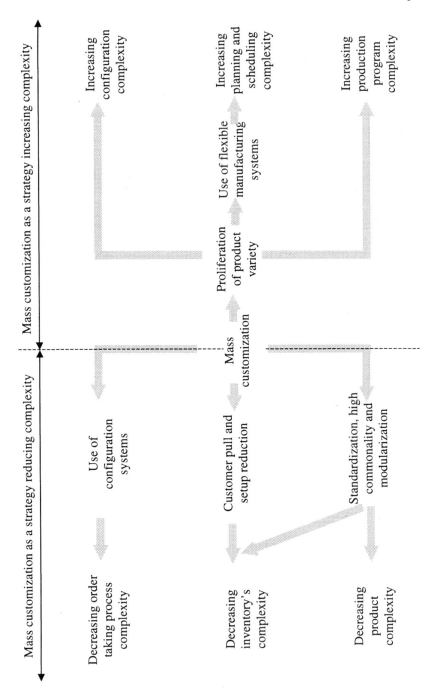

Figure 3-8. Framework presenting the interdependencies between mass customization and complexity

(Source: Blecker et al. 2004e, p. 11)

6.   SUMMARY

In this chapter, the main interdependencies between mass customization and complexity are highlighted. The complexity analysis is focused on the system that consists of three subsystems, namely the product configuration system, manufacturing system and product arrangement system. It is argued that the considered system is the most complex one in a mass customizing enterprise. In addition, it is noteworthy that in order to mitigate high environment complexity characterized by e.g. complex market structures ("markets of one") and rapidly changing customers' requirements, the considered system cannot be a simple one. In the complexity taxonomy of Ulrich/Probst (1995, p. 61) the system will be assigned to the category of extremely complex systems because of high structural variety and high dynamics. Therefore, a certain complexity is required and must be accepted during the pursuit of mass customization.

An analysis is carried out in order to examine as to how mass customization increases complexity. An immediate effect of mass customization is high product variety that triggers high production program complexity, as well as high configuration complexity for customers who are considered to be a subsystem of the enterprise system. Furthermore, the production of a large variety cannot be efficiently realized by mass production systems with high setup times. Mass customization obligatorily assumes the implementation of flexible manufacturing systems on the shop floor. Although these modern systems have a considerable potential to improve manufacturing performance, they increase planning and scheduling complexity.

However, mass customization is a strategy with some potential to reduce complexity. This perspective of viewing mass customization is relevant and must be emphasized. Within the scope of the considered system, mass customization reduces complexity at three main levels which are: order taking process, product and inventory. For example, the competitive success of Dell as a computer manufacturer is basically assigned to the decreasing complexity aspects (complexity breakers) provided by mass customization, namely standardization, low inventory, and direct order taking process from customers. As a result, mass customization does not seem to be an oxymoron with no perspective for success. In some industrial fields, it is even a unique way to outpace competitors.

In parts two and three of this book, we develop solutions which basically focus on information systems and management tools for coping with the complexity drivers in mass customization. But before that, we will introduce in the next chapter a model that enables us to better understand customers' requirements in mass customization. It is argued in many research contributions that the pursuit of this new management paradigm necessarily assumes an optimal comprehension of the customer. On the other hand, the model will provide us with interesting clues in order to better understand complexity in mass customization.

Chapter 4

A CUSTOMERS' NEEDS MODEL FOR MASS CUSTOMIZATION

Up to now, the focus in mass customization research has been mainly oriented toward the product (Svensson/Jensen 2001, p. 5). The development of a high number of product variants that can be manufactured by achieving a near mass production efficiency is mostly the main concern. However, as aforementioned in the previous chapters, customers represent a decisive factor for success in mass customization. They play a very specific role in the value chain and provide valuable input for design and manufacturing. In order to successfully achieve mass customization, customer requirements have to be carefully addressed and understood. For this reason, the main objective of this chapter is to present and discuss a model that enables one to better understand customers' needs in mass customization.

1.   THE CUSTOMERS' NEEDS MODEL

1.1   Information Supply and Need Model

The customers' needs model in mass customization will be elaborated according to the information need and supply model described by figure 4-1. This model stems from the information theory and makes the distinction between the objective information need, subjective information need and information supply.

The objective information need defines the type and quantity of information that a decision maker should use for an optimal achievement of a specific task. The subjective information need represents the information that the decision maker considers to be relevant for completing the task. After determining which information is needed, a corresponding information supply has to occur. It is noteworthy that there are some discrepancies between the objective information need, the subjective information need and the information supply. Therefore, the corresponding circles in figure 4-1 do not superpose. For the problem solving, only a portion of the subjective information need will actually be asked for. This corresponds to the information demand. The intersection of the objective information need, information supply and information demand corresponds to the current information level. This area represents the supplied information that actually works towards the task completion (Wigand et al. 1997, p. 88).

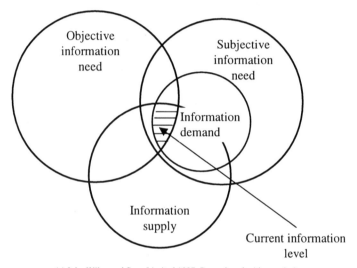

Figure 4-1. Information supply and need model

(Source: Wigand et al. 1997, p. 89)

1.2   The Objective and Subjective Customers' Needs Model

In mass customization, communication between the customers and supplier is necessary. Customers express their individual needs, which enables the mass customizer to manufacture the custom-made product. To relate customers' needs to the information need and supply model, customers' needs are considered from two perspectives, namely as the information the customers should know or actually know about their own needs. The analogies required to construct the costumers' needs model are represented by figure 4-2.

Information supply and need model	Customer needs model
Decision maker	Customers
Task	Product selection task
Objective information need	The information that the customers would require to know about their needs in order to be able to select an *optimal* product variant.
Subjective information need	The information that the customers have identified as necessary to know about their needs for the selection of a product variant.
Information supply	The information that the customers receive in order to carry out a selection of a product variant (achievement potential of the mass customizer)

Figure 4-2. Required analogies to establish the customers' needs model

The decision makers are the customers. The task that must be fulfilled by the customers is the selection of the optimal product variant from the solution space (achievement potential) of the mass customizer. According to the model from the information theory and the analogies, two main categories of information emerge: the objective information about individual needs that the customers would require to select the optimal product variant and the subjective information about individual needs that the customers actually use to select a product variant. We call the first category of information the *objective customers' needs* and the second category the *subjective customers' needs*. The information supply relates to the information that the customers receive from the mass customizer in order to carry out product selection. This information refers to the achievement potential and is called *offered variety* (Figure 4-3). The model also suggests some discrepancies between the objective needs, subjective needs, and offered variety.

Up to now, the methods used to capture customers' needs generally assume that customers are perfectly aware of their requirements. "Most marketers assume that customers know what they want" (Riquelme 2001, p. 1). For example, customer interviews or conjoint-analysis are based on information that is explicitly expressed by customers. In other words, the supplier captures customers' needs as they are identified and understood by the customers themselves. For this reason, it can be stated that the main focus of the methods used, is the detection of the subjective customers' needs, on which basis the suppliers adjust their offered variety.

As a result, both circles, namely the subjective needs and the offered variety are already known by marketers. The subjective needs can be detected by using relatively easy methods. However, the theoretical model which stems from the information theory allows us expect the existence of a third circle which is referred to as the objective customers' needs. In the technical literature this category of needs is not explicitly mentioned. Methods that can evaluate whether customers would make optimal choices, do not exist. There are just *some indications* that corroborate the idea of the objective needs and the unawareness of customers about what they really want. Jugel (2003, p. 414) points out that a customer needs analysis often results in customers actually preferring a product other than what they normally would choose. Zipkin (2001, p. 82) mentions that customers often have trouble deciding what they want. More commonly, they are unsure about their needs and are not able to clearly articulate them. An empirical investigation of Riquelme (2001, pp. 441) suggests that even if customers have good knowledge about the product, they would not be able to perfectly predict their choices. The customer's own evaluation of product attributes may lead to the expectation that the customer would buy a product other than what he or she actually buys. In order to avoid communication problems, Ulrich/Eppinger (2000, p. 63) suggest that "[w]atching customers use an existing product or perform a task for which a new product is intended can reveal important details about customer needs." Moreover, customers are generally unaware of their requirements until they see them violated.

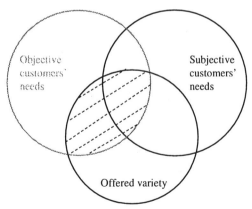

        _ _ _ _   Offered variety corresponding to
                   the objective customers' needs

Figure 4-3. The objective and subjective customers' needs model

(Source: Blecker et al. 2003b, p. 5)

By recapitulating the results attained above, *we define the subjective customers' needs as the individually realized and articulated requirements, whereas the objective needs as the real ones perceived by a fictive neutral perspective. Using knowledge management terms, we argue that the subjective needs are explicit, while the objective needs are implicit* (Blecker et al. 2003b, p. 4).

More simply formulated, the verbally expressed needs that customers *believe* to yield a satisfying product variant are the subjective needs. Suppose that a mass customizer manufactures and delivers to the customer a product variant which perfectly corresponds to his or her expressed requirements. After receiving and using the product, the customer may realize that in fact the product is not the optimal solution which would be more satisfying (von Hippel 2001). In this case the product variant just corresponds to the subjective needs of the customer. However, the objective needs are those that the customer would like to actually be fulfilled, although the customer may not realize them because of e.g. lack of product expertise. The model illustrates that the objective and subjective needs can overlap but do not perfectly superpose. Insofar as the objective and the subjective needs are not identical, the fulfillment of the subjective customers' needs leads to a sub-optimal customer satisfaction, whereas matching the objective needs yields an optimal customer satisfaction.

One may suggest that the established model would also be suitable for strategies other than mass customization. For instance, in mass production customers do not look for individualization, but rather just for affordable low-priced products. Standard products are not conceived to fulfill the needs of a specific customer but those of a hypothetical average customer. Therefore, it makes no sense to speak of the objective and subjective needs. Consequently, the model is not relevant for mass production. In mass customization, customers accept a premium price and longer delivery times just because the product is tailored to suit their individual requirements. If the customized product does not exactly correspond to the objective needs, then customers may doubt the advantages of mass customization and this can have negative consequences on success. Therefore, we argue that the objective and subjective needs' model is only applicable in the case of mass customization.

A comparison of the information supply and need model to the adapted model of figure 4-3 reveals that the circle representing the information demand has no correspondence in the customers needs' model. In fact, it is assumed that the information demand and the subjective need entirely superpose, which means that the totality of the information about subjective customers' needs can be recognized without information loss. This is legitimate because it is argued that the subjective needs are explicitly expressed by customers.

The model involves customers' needs on the one hand and offered variety on the other hand. This may suggest some incoherence or incompatibility in the model. But in mass customization a product is generally considered to be a solution for a specific customer need. Thus, a specific product variant corresponds to a particular customer need and vice versa. For this reason, the terms "customer needs" and "product variant" can be used interchangeably. In the model, both terms are only utilized in order to emphasize the customer's perspective when speaking about "needs" and the mass customizer's perspective when speaking about "offered variety".

1.3   Explanation of the Discrepancies Between the Objective Needs, the Subjective Needs and Offered Variety

There are some discrepancies between the objective customers' needs, the subjective customers' needs and the offered variety. Figure 4-4 outlines the main reasons triggering these gaps.

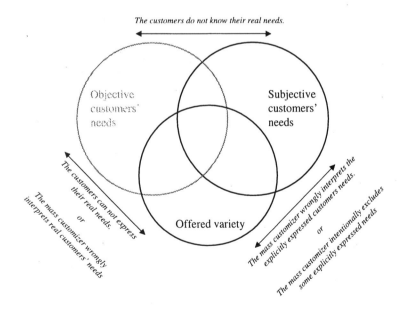

Figure 4-4. Reasons explaining the discrepancies between the objective needs, subjective needs and offered variety

The discrepancies existing between the objective and subjective customers' needs are mainly ascribed to the fact that customers do not know their real needs. The gap between the objective needs and offered variety is mainly due because customers cannot express their real needs and/or the mass customizer wrongly interprets real customers' requirements. Note that a wrong interpretation of real customers' requirements can arise only if there are customers who are able to correctly express their objective needs. If the objective and subjective customers' needs are completely disjoined, this would suggest that no customer is able to identify his or her real needs. Consequently, the discrepancy between the offered variety and the objective needs is exclusively ascribed to the fact that customers cannot express their real needs. This is trivial because if customers are not able to identify their real needs, they cannot express them.

The subjective needs and the offered variety are not congruent because the mass customizer may wrongly interpret the explicitly expressed requirements and/or because of a conscious exclusion of some explicitly expressed customers' needs (e.g. due to economical reasons). The case when the mass customizer intentionally excludes some requirements will not be further discussed here. This case is not problematical because the needs are actually captured, but consciously excluded. In the following, we discuss the main reasons that trigger discrepancies between the objective needs, subjective needs and offered variety. They represent important problems in mass customization that need to be carefully tackled:

- *The customers do not know their real needs.*

Mass customized products are not just products for experts, who know the product configuration that would optimally fulfill their requirements. Mass customization rather addresses a wide range of customers that also includes non-expert customers. These customers often experience difficulties when expressing their own preferences. For example, one can imagine the process a home-owner goes through when designing a customized kitchen (Kahn 1998, p. 28).

- *The customers cannot correctly express their real needs.*

Even if the customers know their real needs, they may have problems to communicate them to others properly. Because the majority of customers do not have the necessary technical expertise, they will not use technical parameters to describe their needs. Rather, they will use verbal language in terms of verbs and adjectives or body language such as gestures or a system of symbols such as pictures and signs. Furthermore, some aspects such as feelings and emotions are difficult to be explicitly expressed.

- *The mass customizer wrongly interprets real/explicitly expressed requirements.*

This aspect can be explained by using the concept addressing the levels of semiotics that assumes three levels of information transmission between a sender and a receiver. The first level is the syntactic level and deals with the transmission of symbols. The semantic level builds upon the first level and considers an additional aspect related to the meaning of symbols. The third level is the pragmatic level including, in addition to the transmission and meaning of symbols, the intention of the sender (Reichwald 1993, p. 451). By applying this model it can be concluded that a disturbance at one of these three levels will lead to a communication problem between the mass customizer and the customer. This can trigger major discrepancies between the different types of needs and the offered variety. For example, at the pragmatic level the mass customizer can consider an important message of the customer to be not relevant.

Because of the discrepancies existing between the objective, the subjective customers' needs and the offered variety, we identify seven regions which are shown in figure 4-5:

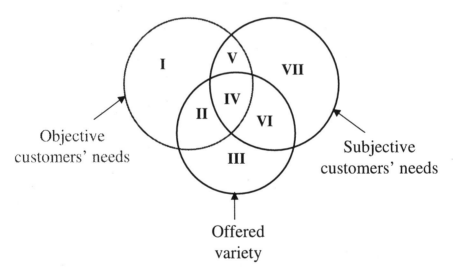

Figure 4-5. Identified regions because of the discrepancies between the objective and the subjective customers' needs

(Source: Blecker et al. 2004a, p. 2)

- Region (I): The fulfillment of the customers' needs situated in this region would generate optimal customer satisfaction. But the corresponding variants are currently not offered by the mass customizer. In comparison to the subjective customers' needs, the objective needs of this region are difficult to detect with the currently available methods and techniques because of their implicit character.

- Region (II): The objective customers' needs in this region can be fulfilled by product variants, which would generate optimal customer satisfaction.
- Region (III): The variants of this region are over engineered, not appreciated by customers and do not fulfill any type of customers' needs.
- Region (IV): The variants of this region simultaneously correspond to the objective and subjective customers' needs and are actually offered by the mass customizer. In this region, the customers are able to select from the solution space of the mass customizer the product variants that will exactly meet their needs and generate optimal satisfaction. This region is the most optimal one under all other regions.
- Region (V): In opposition to the product variants of region (IV) the corresponding needs cannot be fulfilled by the current offer of the mass customizer.
- Region (VI): The customers' needs in this region can be currently fulfilled by product variants which would only lead to a sub-optimal customer satisfaction. A product variant that would better fulfill customer requirements may exist in the solution space of the mass customizer and is situated in region (II).
- Region (VII): The needs in this region correspond to the subjective customers' needs. The variants fulfilling these needs are not offered by the mass customizer. Since these needs are explicit, they can be detected by using methods and techniques such as e.g. customer interviews. The fact that the variants fulfilling these needs are excluded from the production program may be due to a conscious decision from the mass customizer.

The analysis of the discrepancies reveals that there are three challenges that the mass customizer has to face, namely:

1. How to orientate a product design on objective customers' needs in order to reduce the surface of region (I) and thus, to fill the gap between the offered variety and the objective customers' needs.
2. How to rationalize the production program in order to reduce the surface of region (III) by eliminating over engineered variants.
3. How to help customers recognize their objective needs. This means in essence, how to help customers with subjective needs in region (VI) or (VII) in order to buy products in region (II) or (IV) that better fulfill their requirements.

The first challenge relates to a multidisciplinary problem which requires solution approaches and competences especially from the fields of business administration, artificial intelligence, computational technology and consumer psychology. The second one rather refers to business administration and requires methods and techniques to rationalize the production program in mass customization. The third challenge refers to the implementation of adequate customer configuration systems (Blecker et al. 2003b, p. 14).

2.   APPROACHES TO OPTIMIZE VARIETY USING THE CUSTOMERS' NEEDS MODEL

Using the customers' needs model established above we can conclude that there are two directions (1) and (2) as represented in figure 4-6 that the mass customizer has to consider in order to approach the customers objective needs. The objective needs are more important than the subjective needs because they are the real ones that customers actually want to be fulfilled.

The first direction (1) in figure 4-6 deals with how to help customers get to know their real needs better. A problematical situation arises when the customer believes that a variant (A) would fit his or her needs, but in fact there is another variant (B) in the product assortment that would better correspond to his or her requirements. Since the offered variety in mass customization is very large, the customer may not recognize variant (B) and chooses variant (A). In a real shopping environment, the task of making customers recognize their real needs better is generally carried out by advisors, who, due to their experience, are able to find the optimal choice for the customer. However, in a virtual environment over the World Wide Web, the configuration system should offer great potential in order to drive the customers towards the ideal variant while minimizing searching efforts.

But it is conceivable that the objective customers' needs would be fulfilled by no variant in the product assortment. In this case even a good configuration system will not be able to lead the customers to make their optimal choice. Direction (2) in figure 4-6 indicates that the mass customizer has to continuously update the product assortment to approach the objective customers' needs by reducing the over-engineered variants and those corresponding only to the subjective needs. Thus, the main goal is to optimize the overlapping surface between the offered variety and the objective needs circles as shown by direction (2) which represents the right direction for optimizing the offered variety in mass customization.

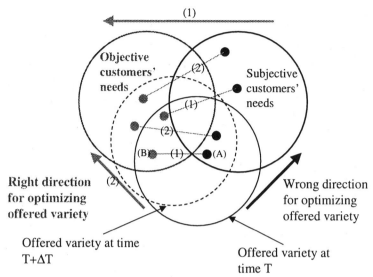

Offered variety at time
T+ΔT

Offered variety at
time T

●   Variant corresponding to subjective customers' needs

●   Variant corresponding to objective customers' needs

(1)   Making customers more aware of their objective needs within the scope of the offered variety

(2)   Optimization of the offered variety in regard to the objective needs of the customers

Figure 4-6. Variety optimization with regard to the objective and subjective customers' needs (Source: Blecker et al. 2003b, p. 15)

3.   SUMMARY

The basis of the objective and subjective customers' needs model in mass customization is the information supply and need model from the information theory. The subjective customers' needs are defined as the individually realized and articulated requirements, whereas the objective needs are the real ones perceived by a fictive neutral perspective. We argue that the subjective needs are explicit and lead to a sub-optimal customer satisfaction, while the objective needs are implicit and yield optimal customer satisfaction.

The model from the information theory allows us expect the existence of a category of implicit needs called the objective customers' needs. However, marketers generally assume the awareness of customers of their needs as well as their capability to explicitly express them. As a result, marketers only capture the subjective needs by means of existing methods such as customer interviews or conjoint analysis. Although in the technical literature, the objective needs are not explicitly mentioned, some indications confirm this idea. Therefore, the model is well-founded and is supported by the information theory as well as by suggestions from business administration.

In mass customization, the distinction between both categories of needs makes sense because nearly everyone should find the product that meets their *objective* requirements. The discrepancies between the objective and subjective needs are ascribed to three main reasons, namely (a) when the customers do not know their real needs, (b) when the customers cannot correctly express their real needs, and/or (c) when the mass customizer wrongly interprets the customers' requirements.

In figure 4-7, the implications of the objective and subjective customers' needs model are summarized. Furthermore, the connections to the results attained in chapter two with respect to complexity are presented. It should be noticed that the increasing complexity in mass customization is essentially ascribed to the proliferation of variety. Therefore, it is important to determine the variety that should be retained and the variety that has to be eliminated. The objective and subjective customers' needs model provides us with relevant clues in order to make an optimal decision.

The model suggests the elimination of the over engineered variants that correspond neither to the subjective nor to the objective needs. These variants are superfluous because they are not perceived by customers at all. The elimination of these variants would trigger a decreasing complexity of the production program and manufacturing.

Another important result deduced from the model is that the product variants just corresponding to the subjective customers' needs have to be eliminated. These product variants are especially problematical because they may confuse customers who may believe that a variant corresponding to the subjective needs is optimal, but in fact there is another product variant in the solution space that fulfills the objective requirements. The elimination of these variants would trigger not only decreasing production program and manufacturing complexities but also a decreasing of configuration complexity from customer's perspective.

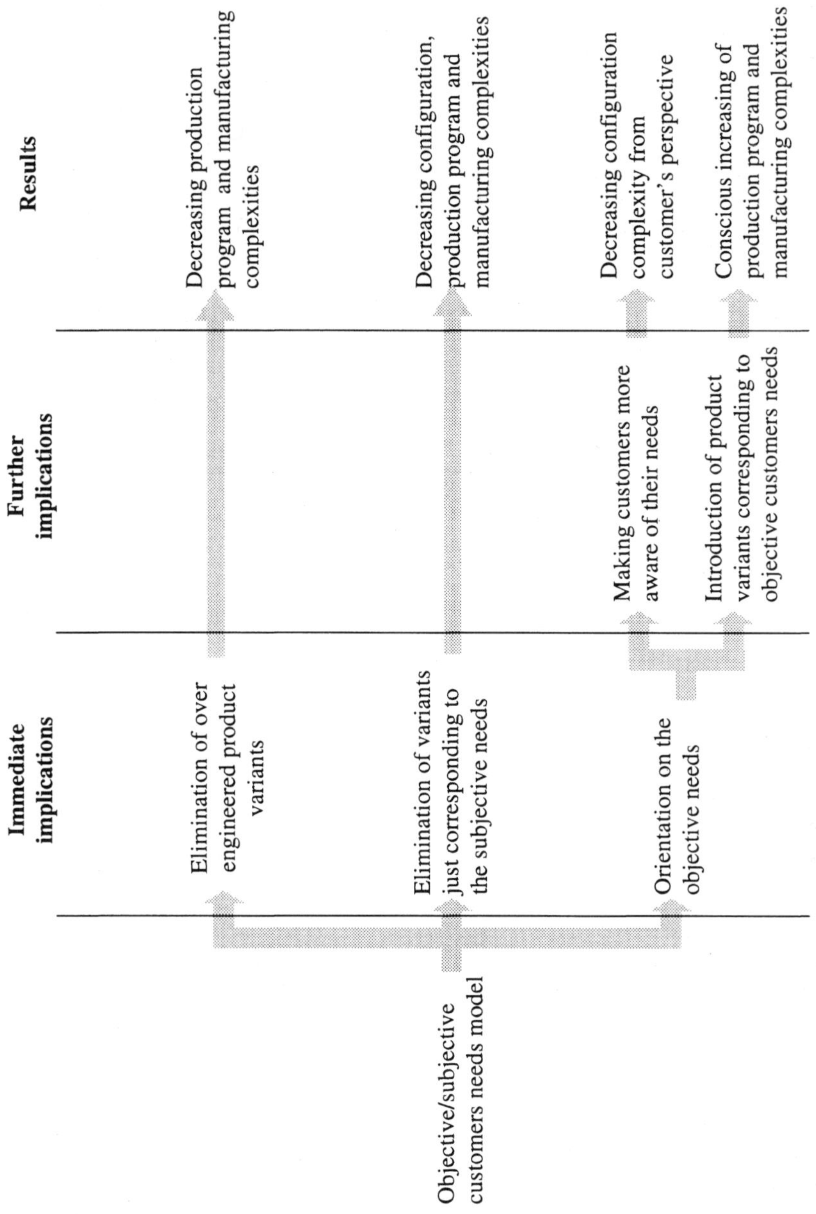

Figure 4-7. Implications on complexity according to the objective and subjective customers' needs model

In order to approach the objective needs, the model suggests making customers better recognize their real needs. In so doing, configuration complexity from customer's perspective can considerably decrease. The model also proposes that the mass customizer should introduce new product variants in order to match unfulfilled objective customers' needs. The decision of introducing additional variety is generally associated with an increase of complexity especially in the production program and manufacturing. However, the complexity that is added in this case can be referred to as a "good complexity" since it is not only consciously introduced, but also valued by customers.

PART II

AN IT INFRASTRUCTURE FOR EFFECTIVE AND EFFICIENT PRODUCT CUSTOMIZATION

Chapter 5

CUSTOMER ORIENTED INTERACTION SYSTEMS

Product configuration systems are considered to be the most important enablers of the mass customization strategy. They are information systems which support the acquisition of the customers' requirements while automating the order taking process. They allow customers to configure their products by specifying their technical requirements. Product configuration systems have an additional relevance because they are one of the few information systems with which customers can directly interact (Bramham/MacCarthy 2003, p. 2).

Up to now, the product configuration process is very technical oriented and necessitates product expertise of the customer. During this process, the objective customers' needs are not taken into account. That is why it is necessary to enhance product configuration systems in order to help customers meet an optimal choice.

In this chapter, we explain the state of the art product configuration systems and classify them in a morphological box. We also outline the shortcomings of the existing systems in the context of mass customization. Subsequently, we introduce the notion of advisory systems which extend the product configuration systems with the main objective to tackle the identified problems that arise when we make the distinction between the objective and subjective customers' needs. After the description of advisory quality and an overview of existing techniques for customer advisory, we depict the requirements for a customer advisory system for mass customization. Then, we describe the technical implementation of a basic advisory system. However, this has to be extended in order to optimally elicit the customers' objective needs. For this reason, suitable levers and technologies are identified in order to provide a conceptual extension of the basic advisory system.

1. CONFIGURATION SYSTEMS: STATE OF THE ART

Product configuration systems or configurators are important enablers of the mass customization paradigm. They are considered to be among the most successful applications of artificial intelligence technology (Felfernig et al. 2002, p. 3). Configurators are information tools that can allow the automation of the order taking process by capturing customers' requirements without involving human intermediaries.

In general, a configurator is implemented at the interface between a supplier and its customers over the Internet. Its principle task is to support customers in the self-configuration of their products according to particular individual requirements. For example, customers can be provided with the possibility to alter a basic product and also to graphically visualize the effects of these changes.

Configurators support the configuration process that requires one to accurately understand the customer's needs in order to create a complete description of a product variant that suits his or her individual requirements. Given a set of customer requirements and a product family description, the task of configuration is to find a valid and completely specified product instance among all of the alternatives that the generic structure describes (Sabin/Weigel 1998, p. 43).

1.1   Different Conceptualizations of Product Configurators

Product configurators have been employed in one form or another for many years. Freuder (1998, p. 29) notes that Lucent technologies has used product configurators for more than 20 years. The main role of configurators is to support the configuration task which is the process of designing a product using a set of predefined components while taking into account a set of restrictions on how the components can be combined (Soininen et al. 1998, p. 357).

In the technical literature, there are many definitions of product configurators. The artificial intelligence community generally addresses a software tool when speaking about configurators. For example, Bourke (2000, p. 2) defines a product configurator as a "...software with logic capabilities to create, maintain, and use electronic product models that allow complete definition of all possible product option and variation combinations, with a minimum of data entries and maintenance". The main technical component of the configurator is the knowledge base which consists of two subcomponents, namely the database and the configuration logic. Whereas the database contains the total set of component types and their instances, the configuration logic specifies the constraints existing between the different components to allow only valid and completely structured product variants. In the following, different classifications of product configurators are proposed, which finally leads to the configurators' morphological box.

1.1.1 Classification according to the configuration knowledge

An important design dimension of configurators is the configuration knowledge which can be based on (a) rule-based, (b) model-based or (c) case-based approaches (Sabin/Weigel 1998, pp. 43). All of these concepts rely on different ontologies. An ontology is defined as a set of concepts to represent, document and store knowledge into a computer or to make inferences (Gruber 1993, p. 1). The different ontologies are required in order to represent the domain knowledge and describe the object types (classes) available in the application domain, as well as the relations among object instances (Sabin/Weigel 1998, p. 43). In other words, ontologies enable a knowledge engineer to map an expert's domain knowledge about the product model into the configuration system's representation, which provides the basis to carry out the configuration task. In the following, a short overview is given of all three paradigms of knowledge conceptualization.

Rule-based approach

Product configuration systems that work on the basis of a rule-based approach are the most commonly implemented in practice (Mailharro 1998, p. 384). The knowledge representation method reposes on rules which have the following form: "if condition then consequence." Therefore, the solutions are derived in a forward-chaining manner. At each step, the configuration system examines the entire set of rules and considers only those that can be executed next. Rule-based systems do not offer a separation between directed relationships and actions. Thus, rules contain both the domain knowledge such as compatibilities, dependencies between components, as well as the control strategy that is necessary to compute the solution to a specific configuration problem (Sabin/Weigel 1998, p. 44). The main drawbacks of rule-based systems are due to the problems encountered during knowledge acquisition, consistency checking and knowledge maintenance, as well as the lack of modularity and adaptability (Guenter/Kuehn 1999, p. 3). A very famous example which illustrates the complexity of knowledge maintenance is XCON. By 1989, the knowledge base of this configuration system consisted of more than 31,000 components and approximately 17,500 rules. The change rate of the knowledge base is about 40% per year (Barker/O'Connor 1989, p. 304). The rules' system represents not only the technical data of components such as performance or quality, but also non-technical data referring to attributes such as e.g., prices that can fluctuate daily. However, these systems are suitable when the solution space of suppliers is characterized by a rigid product structure and low product complexity (Rogoll/Piller 2002, p. 77). For configuration problems with high product complexity, model-based approaches are more convenient.

Model-based approach

The main assumption behind model-based reasoning is the existence of a system's model that consists of decomposable entities and interactions between their elements (Sabin/Weigel 1998, p. 44). Model-based representations can be further classified in logic-based, resource-based and constraint-based approaches.

The most prominent family of logic-based approaches works with description logic, which is a formalism for representing and reasoning with knowledge. The description logic is based on the notions of individuals (objects), concepts (unary predicates, classes), roles (binary relations) and constructors that allow complex concepts and roles to be built from atomic ones. The inference mechanism is based on subsumption.

However, resource-based systems are based upon a producer-consumer model of the configuration task. Each technical entity in the model is characterized by the amount of resources it supplies, uses and consumes. A product configuration is acceptable when resource balancing is realized. This means that the resources that demand different components are each counterbalanced by the resources that the components can maximally supply. The basic schema of the resource's knowledge base consists of a catalog of resource types and a catalog of component types. The main advantage of the resource model is its easy maintenance. However, it is only suitable for products with modular product architectures, which means that products can be constructed on the basis of a few building blocks with predefined interfaces (Juengst/Heinrich 1998, p. 51).

Constraint-based reasoning is a knowledge representation model where components are defined by a set of properties and a set of ports for connecting to other components. Constraints among components restrict the ways various components can be combined to form a valid configuration (Tsang 1993). Configurators that operate with constraint-based reasoning tend to work as follows: the model suggests parts/components from which the user selects one component. The user choice and the restrictions of the product model lead to new configuration possibilities. A restriction can forbid a combination of parts (Part A cannot be combined with part B) or can require a specific combination (Part A requires part C). With each configuration step the product can be more precisely described until no more steps are needed and the product configuration is completely determined (Rogoll/Piller 2002, p. 78).

Case-based approach

The different reasoning techniques that have been described thus far repose on either deductive or abductive schemes to derive solutions. However, the case-based approach takes on a different view and relies on the assumption of claiming that similar problems have similar solutions. The knowledge necessary for reasoning consists of cases that record a set of configurations sold to earlier customers. The control strategy, which is necessary to compute a solution tries to solve the current configuration problem by finding a similar, previously solved problem. Then, the solution to the previous problem is adapted in order to suit the new captured customers' requirements. The basic processing cycle of a case-based reasoning system is: (a) input customer requirements, (b) retrieve a configuration, (c) adapt the configuration to the customer requirements, and finally (d) store the new configuration. The main advantage is the learning capability of the system. Furthermore, in opposition to the other reasoning approaches, no complete product model is needed. However, the implementation of a case-based reasoning system is appropriate when the solution space consists of a few products. As mentioned by Guenter/Kuehn (1999, p. 8), such a configuration system typically leads to conservative solution suggestions that are not innovative.

1.1.2 Classification according to the mass customization strategy

In addition to the configuration knowledge, a relevant criterion for the classification of configurators is the mass customization strategy. As described by Duray et al. (2000, p. 612) mass customization strategies can be classified according to the level of customer involvement and the type of modularity, which leads to the distinction between fabricators, involvers, modularizers and assemblers. With respect to each strategy, the configuration systems should be endowed with different technical capabilities.

Due to a high customer integration level, fabricators have a solution space that theoretically consists of an infinite number of product variants. For example, customers can be provided with the possibility to make a choice among a continuous product dimension range. In other words, the configuration system must be endowed with the technical option that enables the parameterization of component dimensions. Furthermore, a technical integration of the configurator with other information systems, such as Computer Aided Design (CAD) and Product Data Management (PDM) systems may be necessary. That is why from a technical perspective, fabricators would require the most complex configuration systems.

In contrast to fabricators, involvers just give the impression to customers that the product is specifically designed for them. However, the product is assembled on the basis of standard modules and components. Thus, the number of possible product variants is very high but no parameterization of components is necessary. From a technical perspective, the knowledge base of the configurator should enable an efficient administration and maintenance of high product variety.

In comparison to fabricators and involvers, modularizers and assemblers assume the lowest level of customer integration. The set of components and modules is predefined in advance, and the number of product variants is finite and relatively low. Subsequently, the technical complexity of the required configurators is lower than it is in the case of fabricators and involvers. Since modularizers use modularity at an early stage in the production cycle in order to increase commonality, it is advantageous to endow the configurator with a component that supports design engineers in optimizing component sharing among the solution space.

1.1.3   Classification according to organization

For the design of a configurator, it is relevant to make decisions concerning the system's organization which can be either central or distributed. A central configurator works locally within the information system landscape of a manufacturer and its configuration knowledge is completely stored in one single system. All potential product instances that can represent a solution for the customer can be derived from this stored data. In opposition to a central configurator, the knowledge base of a distributed configurator is local, i.e. from a single manufacturer's perspective, incomplete and on its own, it is not capable of providing a comprehensive solution to a specific configuration problem. However, a distributed configurator can be integrated with other configurators (e.g.

suppliers' configurators) in order to form a comprehensive system that satisfies both the customer requirements and the given constraints on legal product constellations for an overall solution. Consequently, a distributed configuration system enables a customer to configure the product by using the configurator's interface of the vendor, whereas in the background, configurators of several suppliers are involved during the configuration task. Such an integration is needed when a central point of configuration knowledge cannot be assumed, e.g. because of confidentiality issues (Felfernig et al. 2002, p. 3).

A distributed configurator has to communicate with other supplier's configurators in order to retrieve relevant data and to generate valid and consistent product instances for customers. Within the scope of the CAWICOMS project (http://www.cawicoms.org/), a prototype for supporting distributed configuration has been developed. The exchange of information between configurators has been technically realized within the JAVA-framework, as well as by means of XML documents. This enables a synchronous communication between configurators, in the case of a web-based online configuration and an asynchronous communication, in the case of a manual configuration carried out by product experts, which includes the possibility to send a mail notification to the customer.

1.1.4   Internal vs. external configurators

Configurators can either be designed for internal or external use. Internal configurators are implemented for the company's internal use, for instance in order to support sales experts in collecting a customer's requirements and translating them into technical features. Forza/Salvador (2002, pp. 87) show through a case study that the implementation of internal configurators to support sales representatives significantly contributes to an increase in the effectiveness and the efficiency with which the company translates the customer's needs into product specific documentation (e.g. bill of materials, routings). As a result, the engineering process becomes faster because the configurator is able to check whether the customer's requirements can be fulfilled with the available solutions. When the product has thus far never been manufactured, the product configurator is able to automatically derive some technical characteristics that are relevant for manufacturing from commercial ones. Thus, internal configurators exculpate e.g. engineers from tasks with no additional value such as the redesign of available solutions and enable them to gain and allocate much more time for development tasks.

Whereas internal configurators support sales representatives or engineers by facilitating and automating certain tasks, external configurators assist customers while configuring their product variants. These configurators are equipped with front-end interfaces that should support customers in selecting the product configurations that best fit their requirements. Due to this very relevant task, Franke/Piller (2002, p. 2) speak of "toolkits" instead of "configurators" because these provide a set of tools that customers can use themselves. Furthermore, in opposition to internal configurators, external configurators can apparently yield additional value to a supplier's web site, which is appreciated by customers.

1.1.5   Classification of configurators according to the nature of interaction

A further classification of configurators can be made according to the nature of interaction which can be either offline or online. Offline configurators work independently from networks. The data being necessary for the configuration task is stored on a data carrier such as a floppy disk, CD-ROM or DVD-ROM. The main advantage of offline configurators is the fast access to a huge amount of data without long download times. However, the main drawback consists in the data up-to-datedness. After configuring the product, customers can send the specifications via e.g., e-mail or fax to the supplier. Consequently, a break in the information flow arises.

Online configurators enable one to avoid the main disadvantage of offline configurators. The configuration knowledge is stored on a central web-server. Therefore, the updating of the knowledge base can be efficiently carried out and all customers who have access to the configurator retrieve effective data. Online configurators can be further divided into two categories that are online configurators with central data processing and online configurators with local data processing. Online configurators with local data processing require the load of the configuration application (Java Applets, Full Java Applications) onto the customer's local unit. However, configurators with central data processing are characterized by continuous communication between the supplier's central unit and the customer's local unit. The main advantage of the configurators with local data processing is the speed because the local unit will provide the required computation performance. The main disadvantages have to do with security risks, as well as the inability to analyze the customer's behavior through data mining because user data cannot be captured during the interaction process. However, configurators with central data processing avoid the main disadvantages of the configurators with local data processing. The main drawback of these systems is essentially ascribed to eventual long waiting-times between different configuration steps. This can disturb a relevant buying criterion, which is the flow-experience.

1.1.6 Classification of configurators according to the updates' execution

A relevant criterion that has to be taken into account when designing configurators is the mode according to which the updates of the knowledge base take place. This classification dimension does not deal with the easiness of the knowledge base maintenance, which rather depends on the configurator reasoning method. However, it addresses another important technical aspect concerning whether the update modes should be push or pull. A push mode is realized when the central unit containing the product configuration logic and situated at the supplier systems communicates product updates to the customer's local unit. In this mode, the central unit imposes the updates that have to be accepted by the local unit. In opposition to the push mode, the pull mode is ensured when the local unit retrieves the updates if required.

1.1.7 Classification of configurators according to the scope of use

With respect to the scope of use, we make the distinction between single-purpose and general-purpose configuration systems. A single-purpose configurator is developed in order to support the sales-delivery process of a product or a set of products of only one company or business field. Single-purpose configurators are also called special-purpose configurators and may be designed for a particular industry such as the window and door industry. Generally, such configurators could not be modified to include products that are substantially different.

However, general-purpose systems are used for the configuration of diverse product types in different industrial fields. Single-purpose configurators are usually programmed in some suitable programming language, while general-purpose systems try to minimize the amount of programming needed for an individual configuration model (Tiihonen/Soininen 1997, p. 16). Nowadays, due to advances in knowledge acquisition, general use configuration systems prevail, primarily in commercial environments.

1.1.8 Classification of configurators according to their complexity

Product configurators can be classified according to their design complexity. Tiihonen/Soininen (1997, pp. 16) distinguish between primitive, interactive and automatic configurators. Primitive configurators are the simplest ones. They merely record the configuration decisions made by the user without ensuring that all of the decisions are valid with respect to each other, or that all of the necessary choices have been made. In contrast to primitive configurators, the interactive ones are capable of checking the validity of configuration decisions, while guiding users in making all of the necessary decisions. In addition to the functionalities of interactive configurators, automatic ones are able to provide full support and to automatically generate, on the basis of customer requirements, parts or even entire configurations. Companies have to analyze their requirements in detail before opting for a configuration system. For certain applications, an interactive configuration system will be sufficient. Therefore, it is not necessary to implement an automatic one which is associated with higher acquisition and implementation costs. Furthermore, the decision which relates to the type of configurator to be used depends on the targeted user groups. Users with a large amount of expertise in the product domain will rather prefer an interactive one in order to configure their products on a technical level by selecting the product options on their own. In this case, the configurator provides support with regard to consistency checks. In contrast, users with little or no product knowledge are probably more effective when the mass customizer implements an automatic product configurator that finds an optimal product on the basis of customers' requirements.

1.1.9   Classification of configurators according to their integration level

The integration level of product configurators in the existing information system landscape of the company is a relevant criterion for the classification of configurators. One can distinguish between stand-alone, data-integrative and application-integrative configurators.

Stand-alone configurators cannot be integrated because they do not dispose of interfaces to other information systems. Furthermore, this class of configurators may lead to data inconsistency. For example, it may happen that engineers change data in the Product Data Management (PDM) system without updating the knowledge base of the configurator. This may lead to a problematical situation when customers order product variants that are no longer available. This drawback can be avoided by implementing a data-integrative configurator. Suppose e.g. that both configurator and PDM system have access to the same database where the product model is stored. An updating action will not lead to inconsistency because both systems retrieve the same data. Moreover, data-integrative configurators avoid data redundancy. However, this type of integration does not indicate the integration of the involved applications. This can be realized when implementing application-integrative configurators. For example, the integration of the configurator and the CAD-system will lead to a relevant advantage when drawings of new parts or components can be generated automatically. Furthermore, the integration with customer relationship management (CRM) systems could help suppliers better supporting their customers. For instance, based on data stored in the CRM system that are gained in earlier configuration sessions of a customer, the supplier can pre-set values for configuration decisions according to the customer's profile. In most industrial cases the interfacing of the configurator with the information system landscape is of high relevance and especially with PDM, CAD and ERP systems, principally in order to ensure data consistency and to avoid redundancies.

1.1.10 Classification of configurators according to the solution searching approach

The solution searching approach is identified as a relevant dimension for the classification of configurators. We distinguish between searching by technical elements or by features. Searching by technical elements means that the configurator enables customers to start from a standard product and then to specify step-by-step product options. However, a configurator working by features provides the possibility for customers to specify their requirements in terms of product functionalities. Then, the configurator performs a search in order to find the product variants in the solution space that best fit the features specified by customers.

The decision which relates to the type of configuration system to use, again principally depends on the targeted user groups. Customers with little or no product expertise would prefer a feature-based configuration, whereas expert users would be more satisfied with a more technical approach. Product experts generally prefer to have the possibility to steer the configuration process based on a technical product description. For instance, in the case of a car, a configuration based on technical elements leads to possible choices concerning the engine (engine type and power), the type of car (hatchback, notchback or limousine) or other technical elements (e.g. anti-lock brake system or an electronic stability program). In contrast, feature-based configuration reposes on a less technical point of view and deals with product features that are expected by the customer, e.g. the purpose of the car (city traffic or highway traffic).

1.1.11 Classification of configurators according to their support of the product life cycle

Supporting the product life cycle relates to product reconfiguration that can be realized by the configurator or not at all. Especially in the field of business-to-business, reconfiguration plays an important role. It is necessary when the customer would like to upgrade the product by including new or better functionalities or to replace non-functioning parts or modules for which identical replacements no longer exist (Sabin/Weigel 1998, p. 47). The different cases that can be encountered are: (a) configurator without a reconfigurator, (b) separate configurator and reconfigurator modules and (c) integrated configurator and reconfigurator. In case (a), the customers are not provided with the possibility to reconfigure their products. In some businesses, this is legitimate. For example, when a customer buys a car, the upgrading possibilities over reconfiguration are not necessary. However, for some investment goods, reconfiguration is necessary. Case (c) is more advantageous than case (b) because the integration of the configurator and the reconfigurator avoids data redundancies.

1.2   Configurators' Morphological Box

In this section, we propose to represent the identified dimensions for the classification of configuration systems in a morphological box. The morphological box was introduced for the first time as an efficient tool for creativity and the structuring of ideas by Zwicky (1966, pp. 114). The main advantage is to present in a straightforward manner all of the possible solution alternatives for a specific problem.

The configurators' morphological box described by figure 5-1 should provide software engineers and developers with the main dimensions to be considered when designing a configurator. The decisions that should be made essentially relate to the determination of the values to be taken by the dimensions. The zigzag line of figure 5-1 shows the relevant characteristics of an example of configurator software with respect to each dimension. It is noteworthy that by taking into account the number of dimensions and their corresponding values, there are $3x4x2x2x3x2x2x3x3x2x3=31104$ possible types of configurators.

As aforementioned, the configurator could have interfaces to customers and it is considered as one of the most important information systems. In the following, we only concentrate on configurators with which customers directly interact during the web-based configuration of product variants. The morphological box shows that the customer perspective is technically not strongly considered. By designing configurators, the technical aspects dealing with how to model the knowledge base or how to adjust a defined strategy, etc. are generally better taken into account than the technical aspects that address the ability to lead customers in a fast-paced manner and with a low amount of effort to their optimal choice.

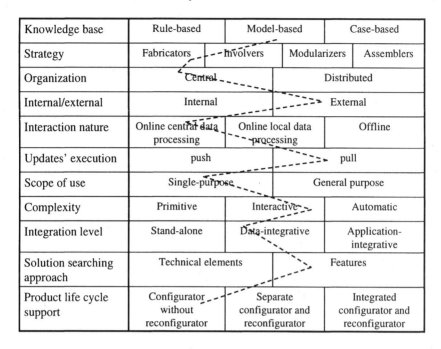

Knowledge base	Rule-based	Model-based	Case-based
Strategy	Fabricators	Involvers	Modularizers	Assemblers
Organization	Central	Distributed
Internal/external	Internal	External
Interaction nature	Online central data processing	Online local data processing	Offline
Updates' execution	push	pull
Scope of use	Single-purpose	General purpose
Complexity	Primitive	Interactive	Automatic
Integration level	Stand-alone	Data-integrative	Application-integrative
Solution searching approach	Technical elements	Features
Product life cycle support	Configurator without reconfigurator	Separate configurator and reconfigurator	Integrated configurator and reconfigurator

Figure 5-1. Morphological box: classification of configurators
(Adapted from: Blecker et al. 2004c, p. 30)

The next section deals with the main shortcomings of product configurators which are considered to be a means to directly interact with customers in mass customization. Driven by the objective to improve the performance of product configuration systems during customer interaction, we point out the necessary requirements for better customer support. The main goal is to find the product configurations that best fit the customers' requirements.

1.3 Shortcomings of Configuration Systems

Compared to standard products, customers face a complex decision making process when buying mass customized products. They have difficulties to decide on their preferences between different alternatives and to compare performance/price ratios of distinct variants. This problem often arises, because in practice the technological perspective generally dominates the user perspective when addressing configuration. It is noteworthy that the existing definitions related to configuration systems, sometimes called configuration toolkits, have a strong technological orientation addressing a software tool, e.g. Bourke (2000, p. 2). In addition, in the technical literature configuration toolkits are criticized more often. Rogoll/Piller (2002, pp. 82) have shown through a market study on configuration systems that there is no standard software solution that is able to fulfill the optimal requirements from the customizer and customer perspectives. Von Hippel (2001, p. 11) criticizes the implemented configuration toolkits in the automobile industry and points out that "...auto makers allow customers to select a range of options for their "custom" cars - but they do not offer the customer a way to learn during the design process or before buying." Learning during the design process means that customers should be provided with the possibility to verify, before placing their buying orders, whether the configured product meets their expectations exactly or not. In the computer industry, von Hippel (2001, p. 12) mentions that websites of computer manufacturers simply provide their customers with the possibility to select components such as processor, chips and disk drives from the available options. Due to the strong product orientation of available configuration toolkits, customers are not able to design the product corresponding to their real expectations (objective needs). It often happens that customers notice that the product is different from what they have expected after receiving or using it.

Furthermore, designers of configurators to a great extent concentrate on the back-end technical aspects. They overlook that especially in the field of business-to-consumer, customers are generally not product experts because they have no technical knowledge about the product and they cannot express their preferences in terms of technical specifications.

The mentioned problems that can occur during interaction can be further explained by the customers' needs model. In effect, the configuration toolkits that are developed thus far only capture the explicit subjective needs. They do not provide customers with the necessary support in order to recognize and then to communicate their objective needs. As a result, the discrepancies existing between the offered variety and the objective customers needs cannot be reduced (see figure 4-6).

Therefore, customers should be adequately assisted during the configuration process. For instance, when buying a car, customers would stipulate the need for the car to maneuver well when parking. In product specific terms, this would mean to endow the car with a servo-steering mechanism. Thus, the language in which customers identify and understand their needs is completely different from the language used by engineers and also product centered configuration toolkits that consists of modules and components. Therefore, configurators must be extended with an additional component that aims at helping customers to better recognize their objective needs.

In the following section, we introduce the notion of advisory systems as a means of customer support in mass customization. These information systems are capable of eliciting objective customers' needs by initiating interactive and personalized dialogs.

2. ADVISORY SYSTEMS AS A MEANS OF CUSTOMER SUPPORT

Unlike in real buying environments where sales experts with adequate experience and knowledge support customers in selecting the optimal product variant, in online environments customers are generally not provided with a good sales assistance. In effect, it is common that the supplier's website only enables customers to make a choice among several product options, either by using a configuration system or directly by selecting products in an online shop. In mass customization, the process of product selection should gain more in importance because customers are integrated in the value chain via the Internet. Online customer assistance is not only a means to reach a competitive advantage by offering an additional value to the customers, but even a necessity. It has to hide the complexity of both the product and the selection process by leading customers to optimal buying decisions.

Advisory systems for mass customization are software systems that guide customers according to their profile and requirements through a "customized" advisory process ending with the generation of product variants which better fulfill their objective needs. They are customer oriented and do not assume any specific technical knowledge of the product. Nevertheless, customers with the necessary knowledge about the product domain also have to be adequately supported.

In our view, the configuration and advisory systems are technically two separate systems. Configuration systems are capable of generating all of the product variants in the solution space of the mass customizer, while advisory systems elicit objective customers needs. On the basis of the captured objective needs, the configurator can generate valid and consistent product variants that are suitable for the customers.

In the following section, the important factors which ensure a high advisory quality in the online channel are proposed. In order to elicit the objective customers' needs, the advisory process has to be customer oriented. It is necessary to support the customer not only in one single buying process, but also during the whole customer-supplier relationship. The availability of online services makes it possible to provide customer advisory round the clock. Therefore, we present the existing alternatives and possibilities for the automation of customer advisory over the Internet, which consequently leads to the special requirements that should be satisfied by on online customer advisory systems.

2.1   Advisory Quality through Customer Orientation

In order to continuously improve the advisory quality in mass customization, the supplier has to maintain a long lasting relationship with its customers. The customer-supplier relationship is initiated when the customer and the supplier come into contact for the first time. When several buying cycles are carried out, one can speak about a long term relationship. The buying cycle includes several phases which run from needs' creation to after sales. The supplier has to understand the different phases of the buying cycle in order to provide comprehensive customer support.

In this context, it should be distinguished between primary and secondary needs. The customer is generally motivated by a primary need which triggers a secondary need for certain products or services. For instance, if the customer's primary need is "mobility", the corresponding secondary need may be e.g. to "finance a car". Primary needs can provide some interesting clues for an optimal advisory process. The supplier should detect the consumer's primary needs in order to adequately adjust the advisory service. Furthermore, during advisory, the objective customer's needs should be recognized in order to be able to make suitable product suggestions.

2.1.1   Customer support within the Customer Buying Cycle

The Customer Buying Cycle (CBC) (Muther 2000, p. 14) originates from the Customer Resource Cycle (Ives/Learmonth 1984, p. 1197). It provides a model that structures the relationship between customers and suppliers according to four phases. The CBC enables the determination of the relevant points of contacts (interfaces) between suppliers and customers. It begins with the identification or the creation of the customer's primary needs. The CBC model assumes that the satisfaction of these needs is ensured by a suitable alternative among many products or services, which leads to the fulfillment of the secondary needs. Whereas the primary needs are implicit, the secondary needs are generally explicit. Therefore, it can be stated that there are some correspondences between the objective and primary needs on the one hand, and the subjective and secondary needs on the other hand. The CBC consists of the stimulation, evaluation, purchasing and after-sales phases (Muther 2000, pp. 15):

1. *Stimulation phase:* In this phase, the potential customers have no defined ideas about their needs. However, they are ready to receive and accept

valuable information. If the supplier correctly addresses the latent primary needs in a personalized way, the likelihood to move to the next phase of the buying cycle increases considerably.

2. *Evaluation phase:* The evaluation phase corresponds to the process that the potential customers go through in order to procure information about available products and services that would satisfy their primary needs. At this phase, the suppliers can actively support the potential customers through individual advisory. Conversations through targeted dialogs enable the human advisors to better understand the needs' situation of the potential customer. Accordingly, personalized recommendations can be suggested. The quality of the recommendation strongly depends on the sales advisor who has to get a grasp on the profile and preferences of the potential customer.

3. *Purchasing phase:* The purchasing phase begins when the customer places the order and it ends after the delivery and payment of the product or service. The supplier should support this phase, so that the entire transaction can be carried out smoothly.

4. *After-Sales Phase:* Within the context of Customer Relationship Management (CRM), the after-sales phase considerably gains in importance. It ensures a high level of customer satisfaction, which enhances the long term relationship. This enables the initiation of a new CBC by using the information gained from the customer during the previous buying cycles. In the subsequent cycles, a more personalized customer advisory can be offered to the customers.

Figure 5-2 depicts the CBC. It is obvious that a customer advisory is especially important during the second phase, in which the customer evaluates different product offers in order to decide what to buy. In this phase, the customer must process the information about the product variants that are available. If customers are overwhelmed by the large variety and cannot make a rational decision, the subsequent phases are necessarily sub-optimal. This negatively influences the customer-supplier relationship and the success of mass customization. To ensure a high level of customer satisfaction as well as an optimal personalized advisory, valuable data must be gained about the customer during all of the four phases without exception. The more information a sales advisor knows about his or her customer, the better the advisory process can be adapted.

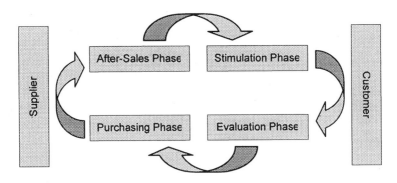

Figure 5-2. The Customer Buying Cycle

(Source: Muther 2000, p. 15)

Target-oriented and personalized dialogs are decisive for the success of the advisory process. They should be adapted to the customers' knowledge level about the product and their preferences. The sales advisors have to ask appropriate questions which relate to the customer's needs situation. On the basis of the answers and reactions of the customer, the course of the dialog has to be continuously adapted by adjusting the following questions. This aims at generating an optimal picture about customers' needs. Another important issue is that the advisor has to check the plausibility of the customer's answers, for instance the compatibility of the requirements and their personal situation.

During the advisory process, the sales advisor has to be appropriately supported, e.g. by information gained about the customer during previous CBCs that are stored in a CRM system. Human advisors should be capable of identifying and understanding the objective customers' needs in order to make suitable product suggestions.

In order to overcome the shortcomings of configuration toolkits, the previously defined advisory systems have to support the customer via the online-channel. Therefore, customer advisory has to be automated and offered as an electronic service.

2.1.2   Customer Advisory as an E-Service

Electronic Services (E-Services) are delivered via the online distribution channel, i.e. the Internet (Bruhn 2002, p. 6). They allow one to electronically integrate the customer into the specification of goods and services within a human-computer-interaction. The potentials that are provided by the Internet such as interactivity, time and location independency, and high download speed enable customers as well as suppliers to gain many advantages. In general, electronic customer advisory is characterized by:

- strong customer integration with a high level of interactivity, and
- an immaterial, very individual result of the service that can be differently appreciated by customers.

In the context of mass customization customer advisory as a lever to improve the usability of configuration toolkits is not a stand-alone service, but a kind of value adding service. In order to use advisory in this case, it is necessary to automate the E-Service. Therefore, it is once again appropriate to consider customer advisory as a dialog between supplier and customer, thus between a sales advisor and a (potential) customer. This communication is direct and mutual, which means that there is a relation between the current dialog and the previous dialog steps. However, these dialogs are not carried out within a face-to-face communication because the conversation is conducted over the Internet. For this reason, one can speak about "electronic customer dialogs" (Meissner 2002, p. 298). Usually, such dialogs are used to lower costs (for instance by reducing personal communication efforts) while improving service (e.g. by time-independent service) at the same time. It is noteworthy that electronic customer dialogs are a pull service, which means that the dialog can be only initiated by the customer.

Customer dialogs can be very complex. That is why not all dialogs can be automated. Furthermore, some kinds of customer dialogs must remain personal within the scope of a face-to-face communication for some reasons, for instance in the fields of premium products where customers generally prefer to be personally advised. Automated customer dialogs can be classified according to the level of standardization and automation. The more standardized a dialog is, the less individualized the questions are that the customers are asked. For instance, Meissner (2002, pp. 301) provides a classification of automated customer dialogs with respect to a decreasing individuality level. On the one hand, suppliers should offer customer advisory round the clock, which requires complete automation, while on the other hand, advisory dialogs should be free and personalized without any restrictions.

2.1.3 Requirements for online advisory systems

Advisory systems have to ensure that the elicitation process is carried out in a personalized way by exploiting information about customers' preferences and knowledge. On the basis of this information, the advisory process can be adequately adapted. The problems that are identified when distinguishing between the subjective and objective customers' needs can be faced partly by simulating the behavior of a human sales advisor. The experts' knowledge should be used in order to derive the objective needs. Therefore, special requirements for an online advisory have to be met:

- *Interactivity*: The electronic dialog must be carried out interactively. It must not be a rigid question-answer-interaction, but a free and flexible dialog. The customer should also be able to steer the dialog. Furthermore, the advisory system has to propose hints or solve conflicts according to the situation (Robben 2001). In addition, the system has to consider the customer's preferences and knowledge interactively by asking questions in a language that can be easily understood by the customer.
- *Dialog sequence*: The dialog sequence should be adapted to the customer who should be only asked relevant questions. Consequently, the sequence of the dialog must be dynamically determined.
- *Presentation of the results*: As in a real advisory dialog, the customer should be provided with explanations about the results. For instance, the advisory system has to explain why some product variants are proposed and some not. Furthermore, the customer should have the possibility to evaluate the quality of the recommendations.

When we consider all of these challenges, it is obvious that the personalization of web pages is the key element of online customer advisory, which means for instance, that the content, presentation or structure of the web page can be adapted to the user (Kobsa et al. 2001, p. 33). In the following section, we analyze some existing systems for customer advisory, namely recommender systems. We categorize these systems according to the nature of their recommendation. Then, we describe their main advantages, disadvantages and analyze if they meet the proposed requirements for customer advisory in mass customization.

2.2   Existing Systems for Customer Advisory

Recommender systems for customer advisory have been successfully implemented in online shops. Generally, they support people who have little or no product knowledge in making a suitable choice (Resnick/Varian 1997, p. 56). The main goal is to provide users with product proposals that would fit their needs. "Recommender systems can be regarded as information agents who serve their users, trying to provide them with the information that best serves their needs" (Paulson/Tzanavari 2002, p. 2). Thus, recommender systems can automate the activities of a human sales advisor and are therefore as a means for the automation of customer advisory.

Schafer et al. (1999, pp. 162) provide a classification of recommender systems. They mainly distinguish between two techniques, namely *personalized* and *non-personalized* systems. Non-personalized recommender systems do not offer individual recommendations that meet the customer's individual interests. Recommendations are rather generated on the basis of an average rating of a wide range of customers who have previously expressed their preferences with respect to the products. For instance, such products that have the highest ratings attributed by other users are suggested to the customers. Furthermore, the suggested recommendations are the same for all different customers. According to the requirements described above for an optimal customer advisory, non-personalized recommender systems cannot adequately support customers during the product configuration process in mass customization. However, personalized recommender systems make product suggestions according to the customers' individual preferences and are good candidates to provide the required customer assistance in mass customization. In this category of recommender systems, it can be distinguished between *attribute based* and *collaborative* systems, which can be respectively referred to as content-based filtering and collaborative filtering.

Content-based filtering is a technique that evaluates the attributes of the offered products with respect to the user model that holds the preferences of the individual customers. It recommends those products whose attributes correspond to the user model. The objects with attributes that are not relevant for the individual customer are excluded from the recommendation. The remaining products to be recommended to the customer are ranked. This method is especially appropriate if customers do have clear ideas about their interests and needs. In addition, it is necessary that the users are able to express their needs properly, for instance in terms of product features, so that the gained data can be directly compared to the product model. Subjective properties such as sportsmanship pose a problem when using content-based filtering. Indeed, product attributes always have to be quantifiable in order to be able to use this technique (Paulson/Tzanavari 2002, p. 2). With regard to the quality of the recommendations, only such products can be found that can be derived from the user model. Therefore, the content-based filtering technique refines product proposals with respect to the same narrow selection of products. This can be disadvantageous because the user does not reach many other products in the product assortment (Schafer et al. 1999, p. 163).

Collaborative filtering generates recommendations on the basis of similarities between customers. In order to determine a product proposal, the customer's profile is related to the profiles of other users. If there are similarities, e.g., in their buying behavior, it is likely that there are also similarities between the users' preferences and interests. Therefore, products bought by other customers can be relevant for a particular user with similar preferences. Data mining algorithms that determine similarity measures are the core of collaborative filtering systems (Paulson/Tzanavari 2002, p. 3). In opposition to content-based filtering, collaborative filtering can also recommend products that are not directly related to the customer's profile. Moreover, it is not necessary to annotate product properties in order to use this technique. Systems which are based on collaborative filtering are easy to be maintained, provided that a high number of users are available (Schafer et al. 1999, p. 162).

Figure 5-3 depicts the differences between content-based and collaborative filtering. It is significant that both techniques do not provide an optimal solution to the problem which arises when we distinguish between the objective and the subjective customers needs. When using content-based filtering, it is assumed that the customers are perfectly aware of their needs, whereas collaborative filtering only exploits similarities between customers who have not necessarily selected products on the basis of their objective needs.

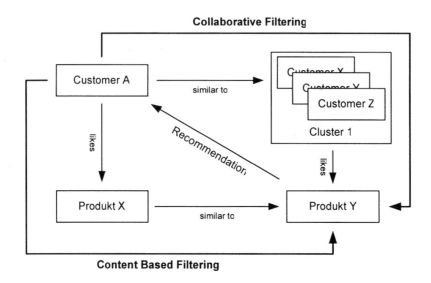

Figure 5-3. Content-based and collaborative filtering
(Source: Klusch 1999, p. 255)

There are other advanced techniques for personalized product recommendations that have been already implemented with success (Burke 2002, p. 331):

- *Demographic Systems*: In addition to the user ratings, these systems collect demographical and social information about the users to determine their preferences.
- *Utility-based Systems*: They use information about features and utility functions over the items that describe the user preferences.
- *Knowledge-based Systems*: They use knowledge about how a particular item meets a particular user's need. On this basis, the system can reason about the relationship between a need and a possible recommendation.

All of the described techniques have their specific advantages and drawbacks. For instance, collaborative techniques are well studied and practical applications have shown the capability of this technique to generate good recommendations that satisfy users. Moreover, because no knowledge about the products is necessary, no knowledge engineering or maintenance activities are required. However, problems arise especially during the start-up phase, since the recommendation relies on a high number of ratings.

Unlike the collaborative filtering, the content-based approach is based on the customer's preferences. If the customers are not aware of their objective needs, then content-based filtering will not lead to optimal results. This technique reposes on questionnaires that just aim at capturing the explicit needs of the customer. Furthermore, with content-based filtering, it is impossible to inspire new customer interests.

However, the knowledge-based approach builds upon expert knowledge and enables the computation of suitable explanations for the recommendation. But the knowledge engineering and maintenance activities of knowledge-based systems are very cost-intensive. In order to overcome the drawbacks of the individual approaches, it is advantageous to implement hybrid approaches which combine several techniques, for instance knowledge-based and collaborative filtering (Burke 2002, p. 336).

Recapitulating all of the results that were attained above, the mere implementation of current recommender systems does not satisfy the criteria and requirements for online advisory systems that are identified in section 2.1.3 of this chapter. Furthermore, they are not appropriate to support customer advisory in mass customization because they are not capable of leading customers to their objective needs. In the following section we propose an advisory system for mass customization that is capable of guiding customers according to their profile and requirements through a "customized" advisory process which ends with the generation of product variants that correspond to their objective needs.

3.   EXTENSION OF THE CONFIGURATION SYSTEM WITH AN ADVISORY COMPONENT

At first, we will describe a stand-alone implementation of an advisory system. This basic system is able to carry out the advisory task and to capture some customers' needs. However, it is not able to optimally solve the problems that are identified when making distinction between the objective and subjective customers' needs. We conclude that the presented approach has to be extended. For this reason, the main levers as well as the required technologies that enable one to better take into account the objective customers' needs are identified. Finally, the technical extensions are integrated into a comprehensive advisory system.

3.1   Basic Structure of an Advisory System

The basic structure of the advisory system should enable the fulfillment of many requirements on advisory systems in mass customization as identified in section 2.1.3. It should work just as is done in real shopping environments by simulating real-world sales advisory dialogs. The simulation of these dialogs with the objective to capture customer requirements ensures a high level of customer orientation. This differs from the application of other concepts such as those based on collaborative filtering which selects products by analyzing the correlation between customers with similar preferences. The advisory system should generate interactive and individualized advisory processes in order to support customers. It should ask targeted questions and provide adequate assistance through alternative answers and explanations. At the end of the process, the product or service alternatives corresponding to the customer needs are displayed. In order to capture customer requirements the system should be able to individualize the dialogs by implementing an advisory logic that adapts e.g. the flow of questions according to e.g. the customer's knowledge level about the product (Blecker et al. 2004a, pp. 4; Jannach 2004, pp. 2).

In order to meet all of the requirements, a knowledge-based approach for advisory and personalization is chosen. This enables sales experts to model the specific knowledge to be used during advisory dialogs. A further advantage of knowledge-based approaches is that they allow the generation of suitable explanations for the customer, which makes it possible to justify the suitability of the set of product alternatives selected for a specific customer (Ardissono et al. 2003, p. 103). In figure 5-4 the architecture of the developed web-based advisory system is pointed out.

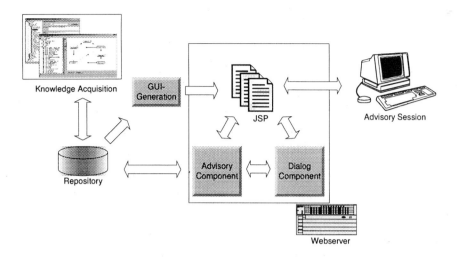

Figure 5-4. Advisory system architecture

(Source: Jannach/Kreutler 2004a, p. 5)

The architecture is in accordance with the Model-View-Controller approach where the interaction control, presentation and data are independently constructed (Krasner/Pope 1988, p. 3):

- The central repository contains the experts' knowledge which consists of the advisory logic, information for personalization, as well as product and customer characteristics. On the basis of the experts' knowledge, the advisory and dialog components are able to carry out the advisory and personalization tasks. In order to reduce knowledge engineering and maintenance costs, a graphical knowledge acquisition component supports product experts in modeling their knowledge (Jannach 2004, p. 2).
- The Graphical User Interface (GUI) is automatically created by a generation module on the basis of the modeled knowledge stored in the repository. Consequently, this makes it possible for domain experts to generate a prototype of the web-based advisory application without programming, which significantly reduces development efforts.
- The users communicate over generated HTML-pages with the dialog component, located on the web server. This component validates and stores user inputs and carries out personalization tasks. For instance, it determines which questions are to be displayed according to the identified product expertise level of the user.
- The dialog component interacts with the advisory component which contains the advisory logic and controls the communication process. In

addition, the advisory component maps customer preferences onto product attributes on the basis of the expert's knowledge stored in the repository.

The system is based on Java technology because Java ensures not only platform independence, but also a uniform framework for the programming of server and client components (Java Server Pages, Java Swing). In the following, we discuss two main system elements which are the dialog and the advisory components.

3.1.1   The Dialog Component

In order to generate goal-oriented and situation-specific dialogs during the advisory process, the experts' knowledge should be made available in an explicit form that can be represented in the system. This requires a simple model for the advisory dialogs in order to ensure a personalized communication between customers and the mass customizer. Furthermore, the model has to support the ability for easy maintenance of the experts' knowledge base. If the dialog model is simple enough and uses terms from the natural language, sales experts themselves will be able to make their knowledge explicit. Consequently, the domain-specific simplified model enables experts to model their knowledge in an easier way than by using the standard modeling methods, such as UML or Petri Nets. Nevertheless, the model offers a formal foundation for the process modeling.

Figure 5-5 describes the developed model with an UML-class diagram consisting of the following components (Jannach/Kreutler 2004b, p. 267):

- The dynamic advisory process consisting of many pages that the user can process step-by-step.
- The pages processed by customers contain, in turn, many questions with predetermined answers in a given layout style, for instance radio buttons, from which customers have to choose. The successive pages of a current page depend on declarative rules which process user inputs.
- The dialog is optionally organized in phases, consisting of a set of pages that provide the user an overview in the progress of the advisory process and also enable direct navigation.
- Each dialog has exception pages, which are activated automatically and independently from the advisory process, in the case when a special event occurs, such as errors or conflicting user inputs. These exception pages interrupt the predefined page flow.

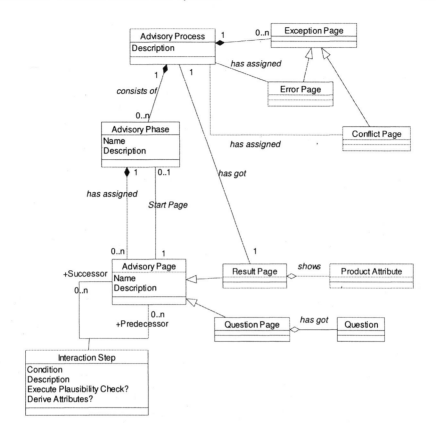

Figure 5-5. Conceptual model for advisory processes

The conception of the advisory process in terms of a pages' flow is ensured by a knowledge acquisition tool that enables the modeling of the real advisory process in a simple manner. The advisory experts can graphically model the advisory process flow by specifying the paths between the advisory steps, as well as the corresponding pages in the form of a tree.

In figure 5-6, the different nodes refer to the advisory pages, whereas the arrows to the conditions which determine the process flow. For example, one condition can be formulated as follows: "choose this path, if the user specifies that he or she is an expert". The specification of these conditions is supported by a context sensitive editor.

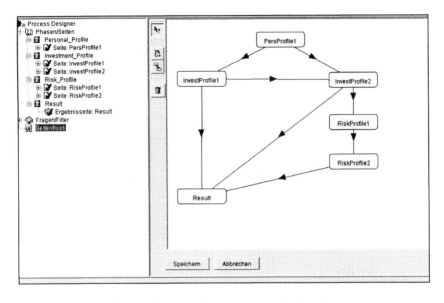

Figure 5-6. Knowledge acquisition component for the advisory process
(Source: Jannach/Kreutler 2004b, p. 268)

Starting from the modeled process, the graphical interface is automatically generated. The interaction can be tabular supported or by a layout containing an avatar. The whole web-application is generated on the basis of the framework containing the dialog and the advisory components, as well as Java Server Pages. These pages can be manually adapted and integrated in already existing websites if necessary. Furthermore, the dialog component enables navigation which refers to the possibility offered to the customer to change the inputs in the different advisory pages when he or she wants. In addition to the dialog component, the advisory process flow is supported by another component, which is the advisory component.

3.1.2   The Advisory Component

The advisory component evaluates customer inputs and poses product propositions by using filter and recommender mechanisms. In the knowledge base the following data are represented (Jannach/Kreutler 2004a, pp. 2):

- Customer characteristics: These are derived from questions which relate to the customers.
- Plausibility tests related to customer answers, which means the rules that are necessary for checking the compatibility of customers' answers. For instance, these tests prevent combinations of answers such as "I am 65

years old" and "I'd like to save money for my retirement" by directing customers to such conflicts.

- Logic for the derivation of customer characteristics: By using a scoring-mechanism, it is possible to draw new characteristics on the basis of the available customer characteristics.
- Product characteristics: All products are characterized by attributes having predetermined value domains.
- Derivation logic for the generation of recommendations on the basis of filter conditions. In these filter rules, customer characteristics and product attributes are used in order to select suitable products among all of the available products.
- Specification of a value utility analysis to rank the proposed products or services as described by Schaefer (2001) and Schuetz/Meyer (2001).

Due to the selected hybrid approach which reposes on knowledge-based and filtering techniques, it is possible to provide customers with personalized recommendations and adequate explanations. This is necessary in order to increase the customers' confidence in the product suggestions of the system (Jannach 2004, p. 2).

Recapitulating, during the advisory process, the advisory component derives new customer attributes and verifies the plausibility of customer inputs. Customers can also request explanations for the questions that are asked, as well as for answer alternatives. At the end of the advisory process, the advisory system proposes product alternatives which are ranked according to their importance for customers by running a value utility analysis. An explanation component explicates on which basis the proposed products are selected.

3.2 Required Extensions of the Basic Advisory System to Better Elicit the Objective Customers' Needs

The presented advisory system does not enable one to optimally cope with the problem that arises when distinguishing between the objective and subjective customers' needs. Indeed, the system reposes on experts' knowledge and recommendation techniques, but it does not take into account some important issues such as e.g. customer behavior. As aforesaid, the objective needs are implicit and cannot be captured on the basis of methods that use explicit knowledge. In order to reach this goal, the described advisory system has to be extended. The main concern is to disable all three faults that may occur during the interaction process (see chapter four, section 1.3).

The identified levers capable of tackling the outlined problems are the dialogs with customers and the mapping techniques which enable the translation of customers' needs into product attributes and vice versa. These levers should be supported by adequate technologies and tools. Whereas the main identified technology is web mining for web data processing, the relevant tools to be implemented are customer interests modeling and web metrics. Both tools aggregate the data provided by web mining in order to present them in an understandable goal-oriented form (Figure 5-7).

Dialogs with customers refer to the communication interface during interaction. Customers should not perceive complexity when specifying their requirements. Therefore, they should only be asked appropriate questions. When customers do not know their real needs, Kansei Engineering and interaction process simplification are suitable solutions, whereas when customers have difficulties in expressing their objective needs, personalization and interaction process simplification are especially relevant. However, the extended dialogs are unsuitable to solve the problem arising when the mass customizer wrongly interprets customer requirements. This is due to three main reasons:

- Dialogs only deal with the front-end aspect between the customer and the interaction system.
- Dialogs do determine the communication process, but not the way information is interpreted.
- Only the available information that is no longer the object of dialogs can be interpreted.

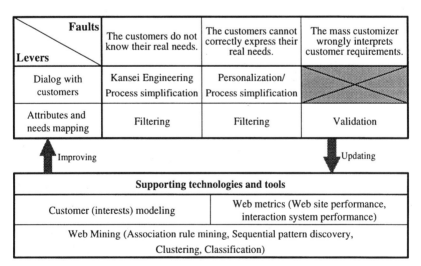

Figure 5-7. Levers and supporting technologies and tools for advisory system extension
(Source: Blecker et al. 2004a, p. 6)

To make customers to be aware of their requirements, a potential solution is to conceive dialogs according to Kansei Engineering which uses verbal language that is close to the language customers are used to understanding. It is a "translating technology of consumer's feeling and image for a product into design elements" (Nagamachi 1995, p. 2). For example, when buying a customized watch, customers are not asked to specify design components, they may be asked e.g. what is the "social position" they want to express.

In addition, in order to avoid overstraining customers during the interaction process, it is relevant that the advisory system guides them to their optimal product configuration by following the shortest path. This refers to interaction process simplification. For example, the advisory system appreciates the customer "knowledgeability" and then accordingly estimates what the technical parameters are that the user is able to specify. If some product parameters are too difficult, the advisory system can set default values without asking questions (Ardissono et al. 2003, p. 107). A superfluous flexibility providing customers with the possibility to specify parameter values that are difficult will confuse rather than help them.

The gained customers' data has to be used for personalization purposes. Personalization aims at recognizing special customer characteristics such as desires and preferences in order to individualize the interaction process. The advisory system should adapt the website layout to customer's requirements and also personalize the formulation of customer dialogs.

The captured customer requirements during the interaction process have to be correctly translated into product specifications. This is ensured by the mapping techniques that not only adequately transform customer preferences and requirements into product specific characteristics (filtering), but also guarantee that the product specifications are adequately mapped to customer needs (validation).

When customers do not know or cannot express their real needs, filtering methods are suitable solutions. Filtering is a collective term for techniques which automatically select product attributes that meet customer profiles and preferences by applying predefined rules, similarities or clustering. For example, with content-based filtering, product configurations can be selected on the basis of correlations existing between product characteristics and a user's preferences which can be captured either implicitly or explicitly. This will considerably restrict the domain of the products' solution space customers would be interested in. It is noteworthy that these filtering methods are based on techniques that aggregate customer data during the CBC, whereas the filtering conditions of the basic advisory system are based on the experts' knowledge.

As opposed to filtering, validation methods have to be implemented to ensure that the mass customizer did not wrongly interpret customer needs.

Thus, the restricted solution space resulting from filtering can be further refined to ensure that the product specifications really correspond to customer requirements. For example, Scheer et al. (2003, p. 12) use a clustering component as a validation method.

The described potential solutions have to be supported by web mining. This technology aims at processing the raw data that is stored in web server logs by applying data mining techniques in order to extract statistical information, cluster users into groups and discover correlations between web pages and user groups (Eirinaki/Vazirgiannis 2003, p. 3). Web mining provides the information necessary to model customer interests in E-commerce and to compute relevant web metrics.

Customer interests modeling is a tool that enables a better understanding of customer preferences and thus, to correspondingly support the personalization of dialogs during the advisory process. In order to better exploit the web mining data, web metrics have to be computed in order to measure the performance of the entire website and especially the performance of the interaction system. For example, it is relevant to appreciate how long one customer has spent on a certain web page or what is the average number of pages that a customer browses before they reach the interaction system. From the web metrics analysis, proposals can be derived for process simplification that can be introduced either automatically online or offline.

The integration of all the proposed solutions in a comprehensive concept will satisfy the requirements placed on advisory systems for mass customization as we defined them in section 2.1.3. With Kansei Engineering and personalized dialogs, customers who have no technical knowledge about products are adequately assisted during their interaction. Furthermore, customer confusion which often leads to sub-optimal product configurations is avoided by process simplification. Due to filtering and validation methods, the advisory system generates optimal product alternatives for customers. Moreover, the advisory systems initiate a virtuous circle which is ensured by a learning process consisting of continuous improvement of the presented solutions and updating of the data processed by web mining techniques. The more customers use the advisory system in mass customization, the better they teach it what they want and the better the advisory system can refine product suggestions leading to better fulfillment of customer requirements.

3.3   Technical Implementation of Advisory System Extensions

We propose to base the personalization task on customer data already existing in a CRM system. It is noteworthy that the CRM system only manages general preferences and properties of customers that are not related to a specific advisory domain. Nonetheless, we can exploit these properties such as the customers' past buying behavior to personalize the interaction flow. As an example we can analyze as to whether the customer is generally interested in low prices or premium products, and we can then parameterize the interaction flow for a specific advisory problem. Personalization based on these bits of information can be carried out on the levels of content and interaction flow (i.e. the complexity level of questions), the style of the presentation (e.g. in form of additional explanations) as well as by proposing default answers that are expected to match the customers' preferences. The overall goal of this personalization lies in simplifying the process and shortening the dialog. In order to exploit the information from the external CRM system, the following functionality has to be implemented:

- The knowledge contained in the CRM system and the advisory system has to be synchronized using defined data exchange interfaces. One possible implementation can for instance be based on a Web Service provided by the CRM system that allows the advisory system to remotely retrieve CRM data over the web using XML as an exchange format.
- The knowledge base in the advisory system has to be extended with business rules that take these additional customer characteristics into account.
- The dialog component has to be extended with the capability to customize the interface according to the customer's preferences and expertise, and for instance to present questions on the adequate level of technicality or even in terms of indirect questions concerning emotions or feelings (Kansei Engineering). Technically, an implementation based on dynamic HTML page-generation such as Java Server Pages allows the system to select an adequate presentation format at runtime according to the current customer profile and the rules defined in the repository.

Regarding the filtering techniques applied, we propose the adoption of a multi-technique approach where the advisory component selects one specific recommendation technique such as collaborative or content-based filtering, depending on the customer's knowledge and the product domain: Collaborative techniques are for instance adequate when the match between customer preferences and products is based on quality and taste. On the other hand, content-based approaches are suitable for customers with sufficient technical knowledge. Finally, case or critique based techniques allow the customers to express their preferences by comparing and rating product alternatives in an intuitive way. In addition to the implementation of these algorithms, the advisory component must implement a rule engine that selects the filtering techniques to be chosen. The interface must also be extended with algorithm specific web pages that for instance allow the users to submit their ratings over HTML.

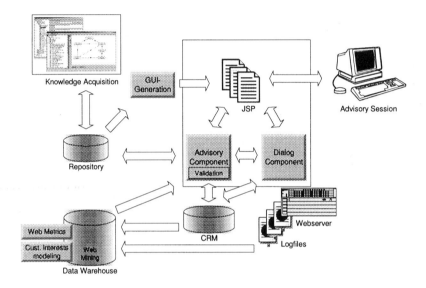

Figure 5-8. Extended advisory system structure for mass customization
(Source: Blecker et al. 2004a, p. 8)

Furthermore, a collaborative approach can incorporate information that is automatically extracted via web mining-techniques from web log-data and user interaction traces possibly stored in an external data warehouse. The result of such a data mining process can be manifold:

- New clusters of users with common preferences, characteristics, and buying behavior can be identified (Customer interests modeling). These clusters can be used by the advisory component to personalize the content for new users by assigning the most probable user class whereby the system can continuously update its beliefs according to user interactions.
- Data on past interactions and buying transactions can be aggregated and new business rules candidates can be inferred (web metrics).

- A validation component attached to the advisory component verifies the proposed products on the basis of behavior analysis. Furthermore, this enables us to check the suitability of the rules of the derivation logic.

Figure 5-8 depicts the overall architecture of the extended advisory system.

4. SUMMARY

In this chapter, we have analyzed the state of the art of configuration systems that mainly determine the customer-supplier-interaction. Several conceptualizations were proposed and represented in a morphological box that shows the main dimensions that must be taken into account when designing a product configurator.

On the basis of the customers' needs model in mass customization that distinguishes between the subjective and objective customers' needs, we were able to outline the shortcomings of the existing configuration systems in the context of mass customization. Up to now, configurators are very product oriented and do not support customers in identifying their objective preferences.

For this reason, we have proposed to extend configuration systems with an advisory component that is able to cope with the described problems. Thus, it is possible to capture the objective needs and to support the customer during the interaction and decision-making process. The main requirements for advisory systems are outlined in order to be able to design such a system. A possible implementation approach is proposed, which starts with a basic system based on a hybrid approach including a knowledge-base and recommender techniques. However, this system is not able to optimally elicit the objective customers needs. Therefore, the main levers and technologies for improving the requirements' elicitation are identified. Finally, the required extensions and technical architecture for an improved objective customers' needs elicitation are proposed.

The developed advisory system only takes into account the complexity faced by customers during the interaction process. It does not support the mass customizer in coping with the variety-induced complexity that is experienced inside the company's operations. In the next chapter, we propose a comprehensive information system that enables one to cope simultaneously with both complexity aspects.

Chapter 6

A MULTI-AGENT SYSTEM FOR COPING WITH VARIETY INDUCED COMPLEXITY

The complexity problems in mass customization are primarily due to the proliferation of product variety. Therefore, variety optimization is a necessary requirement for leading this business strategy to success. The main goal should be to offer an optimal product assortment that consists of product variations that fulfill the objective customer needs by simultaneously minimizing complexity in operations. However, it is not possible to define an absolute optimality level of the product solution space because customer needs and preferences continuously evolve in the course of time. These changing requirements have to be recognized in sufficient time to adequately adapt and update the product assortment.

The discussion of the interdependencies between mass customization and complexity in chapter three has shown two major complexity problems in mass customization that can be classified into internal and external complexity. We define the internal complexity as the complexity that is induced by variety in operations and manufacturing related tasks. In extensive variety environments, customers experience external complexity during the product selection process. In this context, the advisory system only enables one to cope with the external complexity, but does not deal with the internal complexity. Due to changing customer tastes and preferences, it is relevant to determine the product variants to be retained and/or eliminated from the production program. The product assortment optimization problem aims at eliminating the product variants that are not fulfilling the objective customer needs and only retaining those with high relevance for customers. In so doing, the internal complexity is automatically reduced without restraining the product variants that are required for the fulfillment of the objective customers' needs.

The internal complexity is addressed within the scope of *variety steering*, whereas the external complexity problem is solved within *variety formation*. Variety steering refers to which product variants should be retained or eliminated. It aims at solving the optimization problem that balances the customer's and company's perspectives. However, variety formation relates to decisions concerning the product variants that should be instantly formed and offered to a customer by responding in the best way to his or her specific needs and preferences. In this context, it can be stated that the advisory system supports the variety formation task. In the e-commerce-enabled mass customization, variety formation is of high relevance because customers do not want more choice; they just would like the choice that exactly meets their needs (Piller/Ihl 2002, pp. 16). By assuming the existence of a well-defined product assortment, the main objective of variety formation is not to determine what product variants are generally feasible, but to form the product variants with the best chances to fulfill particular customer needs. The suitable product variants are determined by taking into account both customer requirements and the solution space that consists of all of the theoretically possible product variants. As a result, the variety formation task assists customers during the online-buying process in order to lead them to an optimal decision.

Up to now, there is no concept that enables mass customizers to conjointly deal with variety formation and steering tasks. That is why in this chapter, we propose to develop a comprehensive information system that simultaneously addresses both tasks. In order to determine which technology is suitable for the development of such an information system, it is important to first identify the reasons why existing methods and tools for variety formation and steering are not adequate. In effect, it is common that during variety formation, no information is generated about the extent to which the product variants fulfill customer requirements. It is generally supposed that customers are perfectly aware of their requirements. Consequently, the task of forming respectively configuring a suitable product variant out of modules is left to customers. According to the objective and subjective customer needs' model, this strategy is not optimal because customers may be unaware of their real requirements. In addition, state-of-the-art variety steering methods are generally based on centralized approaches. In other words, there are human managers who decide about the retention or elimination of product variants by applying centralized methods such as ABC-analysis, contribution margin accounting, or activity based costing, etc. Centralized approaches have many disadvantages that are due to the following problems:

- *Complexity*: Because of the multitude of product variants, managers are not able to deal with each variant. They rather aggregate data. Consequently, a part of the relevant information for supporting variety steering decisions may become lost.
- *Incomplete information*: In order to reduce the complexity of the product assortment's analysis, managers concentrate on a few aspects of variety steering, e.g. costs' considerations. They base their decisions only on the available information that can be incomplete.
- *Expensiveness of information*: Information, especially from customers, which can be helpful for variety steering, may be very cost-intensive.

To deal with all of the shortcomings of the centralized approaches, the idea is to base both variety formation and steering on a decentralized approach. This can be achieved within the scope of an electronic market mechanism coordinating a multi-agent system. Multi-agent technology enables the decentralization of problem solving tasks by avoiding the shortcomings of centralized approaches. It is a suitable approach in order to simultaneously support variety formation and steering in mass customization. In the next section, we describe the fundamentals of multi-agent systems. Then, we introduce the multi-agent based approach for coping with the variety induced complexity in mass customization.

1. BASICS OF MULTI-AGENT SYSTEMS

The main goal of this section is to provide a short overview on multi-agent systems by drawing on the leading literature in this field. At first, we describe the context in which multi-agent systems are addressed. Then, we separately deal with agents as individual entities and multi-agent systems as social systems.

1.1 (Distributed) Artificial Intelligence

Intelligent (rational) agents are discussed in the context of artificial intelligence (AI). However, multi-agent systems are addressed within the scope of distributed artificial intelligence (DAI). In fact, it is not self-evident to consider artificial intelligence as a discipline that can be exclusively assigned to computer science because it is a multidisciplinary research field. Russell/Norvig (1995, pp. 4) have identified two main dimensions according to which the definitions of artificial intelligence can be classified. Whereas the first dimension is concerned with thought processes and reasoning, the second dimension rather relates to the behavior's perspective (Figure 6-1).

Furthermore, the success of the applications of artificial intelligence is measured in terms of human performance or rationality that is defined as a kind of ideal concept of intelligence.

Systems that think like humans (thought processes/reasoning)	Systems that think rationally (thought processes/reasoning)
Systems that act like humans (behavior)	Systems that act rationally (behavior)

(c) 2003. Adapted by permission of Pearson Education, Inc., Upper Saddle River, NJ.

Figure 6-1. Categories of definitions of artificial intelligence
(Adapted from: Russel/Norvig 1995, p. 5)

Before the emergence of the multi-agent paradigm, knowledge-based and expert systems first began to cope with large problems in relatively large domains. The modularity of knowledge bases and the parallelism of the computation task enable one to find solutions to large problems in a distributed manner. Distributed artificial intelligence is a field that deals with the development and evaluation of different strategies that relate to the distribution of large problems (Kirn 2002, p. 54). It considers intelligence to be a system property. However, artificial intelligence regards intelligence as an individual property. Furthermore, artificial intelligence deals with cognitive processes, whereas distributed artificial intelligence focuses on social processes.

Bond/Gasser (1988, p. 3) classify distributed artificial intelligence into three sub-areas, namely parallel artificial intelligence, distributed problem solving and multi-agent systems. In this classification, parallel artificial intelligence can be considered to be less important than the other areas. Indeed, parallel artificial intelligence deals with the acceleration of the solution-finding process through the parallelization of the problem-solving task. It does not address the distribution of complex problems. However, distributed problem solving aims at dividing a particular problem into a number of modules that cooperate with each other in order to solve a complex problem. Thereby, the manner by which the knowledge is distributed plays an important role. In contrast, multi-agent systems do not focus on how the problem is distributed, but on solving the problem in a bottom-up manner. Agents are considered as nodes that behave according to coordination and cooperation processes in order to find a solution to a specific problem.

The classification proposed by Bond/Gasser (1988) is more often criticized. For instance, Goessinger (2000, p. 85) complains about the logical levels of the used terms. Whereas parallel artificial intelligence needs several problem solving nodes, distributed problem solving can be regarded as a technique that uses parallel artificial intelligence. Furthermore, multi-agent systems define a structure that is implicitly based on the previous concepts. Therefore, this classification does not seem to be unambiguous. Moulin/Chaib-draa (1996, pp. 3) propose a more intuitive classification, in which distributed artificial intelligence is considered to be the intersection between artificial intelligence and distributed computing. With this classification, it is possible to clearly distinguish distributed artificial intelligence, which is a field that includes multi-agent systems and distributed problem solving (Figure 6-2).

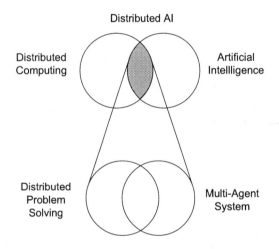

Distributed AI

Distributed Computing

Artificial Intellligence

Distributed Problem Solving

Multi-Agent System

Figure 6-2. Categorization of Distributed Artificial Intelligence
(Source: Moulin/Chaib-draa 1996, p. 4)

It follows that, among other tasks, artificial intelligence deals with the construction of intelligent agents (Wooldridge 2002, pp. 9). More precisely, it focuses on the intelligence component (e.g. planning, learning or picture recognition) that enables an agent to make independent decisions. Nevertheless, the development of rational agents is possible without having to solve all of the research problems in artificial intelligence. In other words, rational agents can be constructed by drawing on specific techniques from software engineering and artificial intelligence. In this context, Russel/Norvig (1995, p. 7) adapt the notion of intelligence in that they consider a rational action as the best possible action. "Acting rationally means acting so as to achieve one's goals, given one's beliefs. An agent is just something that perceives and acts. [...] In this approach, AI is viewed as the study and construction of rational agents."

1.2   Intelligent Agents

Wooldridge (2002, p. 15) defines an agent, as "...a computer system that is situated in some *environment*, and that is capable of *autonomous action* in this environment in order to meet its design objectives." Therefore, an agent "... can be viewed as *perceiving* its environment through *sensors* and *acting* upon that environment through *effectors*" (Russel/Norvig 1995, p. 31). Figure 6-3 illustrates an agent in its environment. The agent takes input from the environment through its sensors and carries out actions that are produced by its effectors.

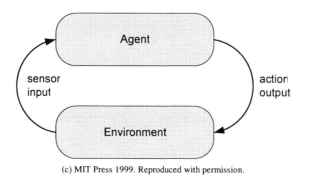

(c) MIT Press 1999. Reproduced with permission.

Figure 6-3. An agent in its environment

(Source: Wooldridge 1999, p. 32)

However, it is not easy to provide a definition for intelligent agents. In effect, Wooldrige (2002, p. 23) argues that the question as to what intelligence is difficult to answer. For this reason, it is more appropriate to list the capabilities that are expected to be satisfied by an intelligent agent. Wooldridge/Jennings (1995, p. 119) make the distinction between weak and strong agents. A weak agent is considered as the most general form of intelligent agents with the following properties:

- *Autonomy*: Agents decide about their actions on their own (Rosenschein 1985).
- *Social Ability*: Agents interact with other agents.
- *Reactivity*: Agents perceive their environment and respond accordingly.
- *Proactiveness*: Agents can take the initiative to carry out an action.

The strong notion considers an agent as a computer system that has – in addition to the properties of a weak agent – some properties that are more similar to human beings. For instance, it is quite common to characterize an agent through mentalistic processes such as knowledge, belief or obligation

(Wooldridge/Jennings 1995, p. 120). For the determination of the agent's inner structure, the agent should be considered in its environment. It takes inputs through its sensors from the environment and produces outputs through effectors. Therefore, the agent needs a program that maps percepts into actions. Russel/Norvig (1995, pp. 40) depict three different architectures for software agents:

- *Simple reflex* agents derive their actions from simple condition-action-rules that build up the connection between sensorial input and actions. Decisions are only made on the basis of the current percept.
- *Goal-based agents* keep track of the environment's state and information about the goal they strive to achieve. On this basis, the agent can apply artificial intelligence techniques for searching or planning in order to derive the most appropriate possible action.
- *Utility-based agents* use a utility function for the generation of a better, more goal-oriented behavior. The utility function maps a state onto a real number that indicates the utility of the current state, in other words, the extent to which an agent is "happy". Furthermore, the agent is able to outweigh the advantages and disadvantages of its available action with respect to the goal it wants to achieve.

Figure 6-4 depicts a possible architecture of a goal-based agent. For the modeling of agents, the Belief-Desire-Intention (BDI) architecture (Bratman et al. 1988; Rao/Georgeff 1991) plays an important role. This model has been derived from the theory of human practical reasoning and is similar to the goal-based approach. A BDI-agent is based on three elements, namely *beliefs* about the environment, *desires* (goals that the agent wants to achieve) and *intentions* (desires that the agent has committed to achieving). The BDI model is particularly interesting because it combines three components: (a) a philosophical component that it is based on the BDI theory of humans' rational action, (b) a software architecture component and (c) a logical component which is a family of logics that capture the key aspects of the model as a set of logical axioms (Wooldridge 2000, p. 7).

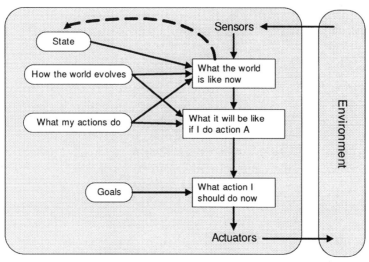

Figure 6-4. Utility-based agents

(Source: Russel/Norvig 1995, p. 44)

In order to completely describe an intelligent agent, it is also necessary to describe the environment to which an agent is tightly coupled. Russel/Norvig (1995, p. 46) differentiate between several properties of environments:

- *Accessible vs. inaccessible:* An environment is accessible when the agent's sensors can obtain complete information about the state of the environment.
- *Deterministic vs. non-deterministic:* An environment is deterministic, if any action has a single, guaranteed effect. This means that the next state is always completely determined by the agent's action given the environment's current state.
- *Episodic vs. non-episodic:* In an episodic environment, the agent's experience is divided into episodes. An episode consists of the agent perceiving and then acting. The quality of an agent's action only depends on the current episode. Subsequent episodes do not depend on the actions that were performed in previous episodes.
- *Static vs. dynamic:* Static environments remain unchanged while an agent is deliberating.
- *Discrete vs. continuous:* An environment is discrete if there is a limited number of distinct and clearly defined actions and percepts.

For the development of intelligent software agents, a kind of abstraction is necessary. Wooldridge (2002, pp. 31) proposes a formalism that provides an abstract view of the agents' environment, the set of actions that the agents can perform and the functions that map percepts into actions. In this model, each agent tries to maximize its expected utility by optimizing the utility function. In section two, we will draw on such a formalism for modeling the multi-agent system for variety formation and steering.

In practice, there are many applications, in which intelligent software agents can be implemented. Nwana/Ndumu (1998, pp. 29) suggest a typology for the classification of existing software agents:

- *Collaborative agents* emphasize autonomy and cooperation with other agents in a multi-agent environment.
- *Interface agents* collaborate with the user. They are a kind of personal assistants which autonomously carry out tasks for users.
- *Mobile agents* are software processes that are not only able to move around networks, but also to interact with foreign hosts and processes.
- *Information/Internet agents* manage, manipulate or collate information from different sources. Such agents support the user in managing a high amount of information over the World Wide Web.
- *Hybrid agents* combine properties of the previously mentioned agents.

Jennings/Wooldridge (1998, pp. 10) provide several examples for the application of intelligent software agents in several domains, for instance industrial applications (e.g. process control, manufacturing or air traffic control), commercial applications (e.g. information management, e-commerce or business process management), medical applications (e.g. patient monitoring), or entertainment. Some of these applications are based on collaborative intelligent agents that are required for the coordination of several autonomous agents. In the following section, we will provide a short overview on multi-agent systems.

1.3   Multi-Agent Systems

"Multi Agent Systems are concerned with coordinating intelligent behavior among a collection of autonomous intelligent agents, how they coordinate their knowledge, goals, skills, and plans jointly to take action or solve problems" (Bond/Gasser 1988, p. 4). According to Weiss (1999, p. 3) a multi-agent system has the following properties:

- Each agent only has a restricted view of the entire problem to be solved. In other words, each agent has to solve a sub-problem.

- The agents' capabilities of solving problems are limited.
- The system is distributed.
- The solution of the sub-problems takes place asynchronously.

   In order to solve complex problems in such a distributed way, agents have to commit to coordination and interaction mechanisms (Fischer et al. 1998). These mechanisms can vary from simple information interchanges to negotiations when an arrangement of interdependent activities is necessary. In most cases, agents act to achieve objectives either on behalf of individuals or as part of some initiative that aims at solving a complex problem. Agents have to interact in accordance with a well-defined organizational context that determines the nature of relationships between agents. However, it is important to consider the temporal extent of relationships because existing relationships can evolve and new ones can be established in the course of time (Jennings 2000, p. 281). Figure 6-5 depicts a multi-agent system with several agents that interact in order to solve a distributed problem. Note that the actions of several agents can interfere in the environment, which consequently increases the demand for coordination.

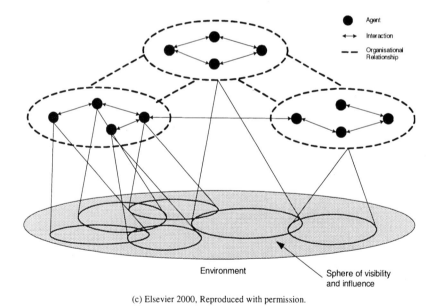

Figure 6-5. A view of a multi-agent system

(Source: Jennings 2000, p. 281)

   Since intelligent agents are autonomous, they cannot be assumed to share the same goal. Similar to human societies, the individual goals of agents can conflict. By decomposing an overall problem into several sub-problems that can be distributed among the agents, each agent can solve the sub-problem according to its individual goal. Then the sub-solutions can be integrated into an overall solution. Notice that Krallmann/Albayrak (2001, p. 120) do not speak about an overall problem to be solved by a multi-agent system. They state that there is neither a global goal to be achieved nor a problem to be solved. Since agents pursue individual goals, a system behavior instead of an overall solution arises.

Wooldridge (2002, pp. 190) identifies task sharing and result sharing according to Smith/Davis (1980) as suitable concepts for solving problems in a distributed and cooperative way. In task sharing, a task is decomposed to smaller sub-problems that are allocated to different agents, while in result sharing, agents supply each other with relevant information, either proactively or reactively. The key problem of task sharing is the allocation of tasks to individual agents. It follows that agents have to reach agreements that are attained within coordination mechanisms. In a literature survey, Corsten/Goessinger (1998, p. 178) identify two main system architectures for the coordination of multi-agent systems, which are: (a) blackboard systems and (b) contract nets.

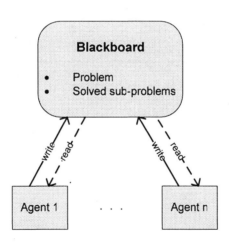

Figure 6-6. Blackboard architecture
(Source: Corsten/Goessinger 1998, p. 178)

Blackboard systems are characterized by two main attributes: Knowledge sources which can be autonomous agents and a shared data structure called a blackboard. The knowledge sources are capable of reading and writing to the data structure. The problem solving proceeds by monitoring the blackboard and writing to it if an agent can contribute to solving a sub-problem. It follows that each agent only implements the communication protocol with the blackboard. A main prerequisite is to cut down the overall problem to several sub-problems at the blackboard (Wooldridge 2002, p. 307). Figure 6-6 depicts the structure of a blackboard system.

The second architecture is contract nets. According to Smith/Davis (1980), Wooldridge (2002, p. 194) defines the contract net protocol as "…a high level protocol for achieving efficient cooperation through task sharing in networks of communication problem solvers." In this architecture, agents communicate with each other in order to place contracts with respect to the demanded and offered services for solving sub-problems. Consequently, the agents can build an electronic market for solving the problem. In opposition to the blackboard system, the contract net architecture requires multi-lateral negotiations between agents. Therefore, agents have to possess the capability of carrying out negotiations in order to reach agreements. Thus, a contract protocol that defines the rules of the communication and negotiation processes has to be implemented (Corsten/Goessinger 1998, p. 180). In a contract net protocol, the negotiation basically proceeds according to the following steps (Wooldridge, 2002, p. 196):

1. An agent recognizes a problem to be solved (*Recognizing the problem*).
2. Therefore, as the manager, the agent announces this problem with the objective to find an agent that is able to solve it (*Task announcement*). This announcement can be a general broadcast to all agents, if the capabilities of the agents are unknown, a limited broadcast to a certain group of agents or a point-to-point announcement.
3. Each agent listens to task announcements and evaluates its appropriateness with respect to task fulfillment. When a suitable task is found, the agent submits a bid to the manager (*Bidding*).
4. The manager receives several bids and makes a decision about the agent that fulfills the task. Subsequently, it sends an award message to the successful bidder (*Awarding the contract*).

The contract net protocol requires a high level of interaction between agents. Therefore, the protocol has to be implemented in a certain language. In multi-agent systems, typed messages on the basis of speech acts' theory are common for information interchange (e.g. Wooldridge, 2002, pp. 164). Based upon this theory, several languages for agent communication such as KQML (Knowledge Query and Manipulation Language, http://www.cs.umbc.edu/kqml), have been developed.

In practice, numerous applications of multi-agent systems have already been implemented with success. In the next section, we propose an innovative application of this technology in order to support dynamic variety formation and steering in mass customization.

2. A MULTI-AGENT BASED APPROACH FOR VARIETY FORMATION AND STEERING

In order to define the multi-agent approach for variety formation and steering, we describe in the following the required assumptions and definitions.

2.1 Assumption and Definitions

The web-based configurators in mass customization are product oriented. They necessitate that customers step by step specify the product variants in technical language. However, the advisory system described in chapter five pursues another logic since it provides customers with full product alternatives that meet their specific profiles and needs. The configuration of products is not carried out by customers themselves, but by the advisory system. The restriction of the huge product solution space to a very few product variants considerably helps customers during the decision making process. This approach will be further pursued due to its relevance in the reduction of the external complexity perceived by customers. At an abstract level, the advisory approach can be modeled as a mechanism that selects suitable subsets of product alternatives to be displayed to customers (Figure 6-7).

If we consider the membership at a subset as a success criterion and that only a few places in the subset are available, then we can imagine that the product variants compete with each other for membership. Thus, a priori, by associating with each product variant an autonomous rational agent, we can construct a multi-agent system, in which agents compete with each other in order to participate in the final subset to be presented to the customer. However, the number of product variants in mass customization can be very high. The assignment of an agent to each product variant would result in a system that can consist of billions of agents. Such a system would not be manageable and is therefore not efficient. Thus, the association of an agent with each product variant is not a promising approach.

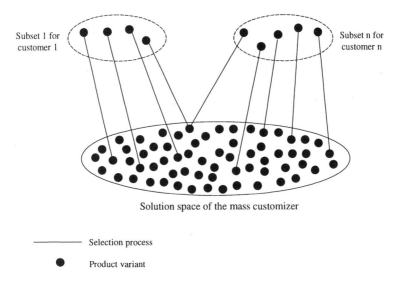

Subset 1 for
customer 1

Subset n for
customer n

Solution space of the mass customizer

———— Selection process

● Product variant

Figure 6-7. The advisory process at a high level of abstraction

However, product variety that is adequate for mass customization has an interesting property. In effect, the best method for achieving mass customization is to develop products around modular architectures (Pine 1993, p. 186). Modules can be defined as building blocks with defined interfaces (Ericsson/Erixon 1999, p. 19). The combination of only a few modules enables the construction of a huge number of product variants from which customers can choose the most suitable configuration. Due to the importance of product modularity that will be dealt with in more detail in chapter eight of this book, we provide the main assumption underlying our approach:

Assumption: Modular product architecture
We assume that the complete set of product variants can be manufactured on the basis of modules.

In order to avoid the difficulty that results from assigning an intelligent agent to each product variant, the idea is to associate with each module variant an intelligent agent. In so doing, the number of agents in the multi-agent population can be kept at a reasonable level, so that the problem remains manageable and the computation resources are not overstrained. Thus, the competition does not occur between the product variants but between the agents that are associated with the module variants. Therefore, we formally provide the first definition:

Definition 1: Module agent

Let M be the set of all modules, $M = \{MC_1, MC_2, \ldots, MC_m\}$. We call MC_i a module class. A module class MC_i contains a set of module variants $MV_{i1}, MV_{i2}, \ldots, MV_{ip(i)}$. p is a function associating an index i of a module class with the index $p(i)$ referring to the number of module variants in a module class. With each module variant MV_{ij}, $j = 1, \ldots, p(i)$ we associate an autonomous rational agent, called a module agent MA_{ij}, $j = 1, \ldots, p(i)$ that disposes of predefined resources and is able to perform tasks.

Modules are classified into must- and can-modules. Must-modules ensure the basic product functionalities, whereas can-modules are optional. According to Ericsson/Erixon (1999), the set of the modules required for the manufacturing of all possible product variants is called a platform. However, we opt for the definition of platforms that is provided by Piller/Warringer (1999) and Wildemann (2003) who understand the product platform as a specific *common module* that is built into a great range of product variants. This definition considers the platform as a module with additional relevance compared to other modules due to its high implementation frequency. In order to make the distinction vis-à-vis other module agents, the corresponding agents are called platform agents. The set of all platform and module agents are grouped in an agent pool.

In order to set the rules of competition between module agents, it is relevant to first define the upper goal that the agents strive to achieve. At this stage, we recall that a main motivation for the development of the multi-agent based information system is to support variety steering tasks. Thus, abstractly, if we suppose the existence of a success criterion, we can say that successful module agents have to be retained, whereas unsuccessful ones have to be eliminated from the agent pool. Therefore, we provide the second definition:

Definition 2: Self-preservation

Each module agent MA_{ij} strives for ensuring its existence by having enough resources to survive.

Definition 2 leads us to consider evolutionary theory (Kauffman 1993), which sees evolution as the result of selection by the environment acting on a population of organisms that compete for resources. The winners of the competition are agents that are most fit for gaining the resources necessary for survival. However, the losers are not able to win resources. The resources of an agent are stored in an account that is defined as follows:

Definition 3: Module agent's account

Each module agent MA_{ij} has revenues and expenses that are summed up in an account Acc_{ij} of monetary units. Acc_{ij} constantly diminishes in the course of time.

The account serves as an internal coordination mechanism for the multi-agent system. The current account surplus refers to the actual resources of an agent MA_{ij}. Each module agent receives a starting amount of monetary units when the corresponding module variant is introduced to the production program. From definitions 2 and 3, it can be concluded that each agent endeavors to maximize its account in order to ensure self-preservation. If the module agent's account becomes zero, then the module agent risks "death", which leads to the elimination of the module variant. The second part of definition 3 reveals that the agent's resources diminish in the course of time even if the agent does not carry out any task. In order to explain what a task is, we provide the following definition:

Definition 4: Module agent's task

The task T_{ij} of a module agent is to form product variants by joining coalitions $C_k, k = 1, \ldots, n$.

The allocation of tasks to groups of agents is necessary when tasks cannot be performed by a single agent. The module agents alone are not able to solve the problem. In our case, each module agent strives to participate in coalitions in order to form product variants with good chances of membership in the final subset that is displayed to the customer. Thus, each agent has to cooperate with other agents in its environment in order to fulfill its task. Thereby, the autonomy principle of agents is preserved because each agent can decide whether to take part or not in a product variant. By forming coalitions, each module agent strives for improving its personal utility/account via cooperation. Module agents follow the economic principle of rationality and attempt to form a coalition that will maximize their own utilities. Module agents contribute differently in the satisfaction of customer requirements. Therefore, they will not be equally efficient. It is expected that some agents are more successful than others.

In order to participate in a coalition, the module agent has to pay a certain fee. In opposition to other work in multi-agent systems (e.g. Shehory/Kraus

1995), the module agent is allowed to participate in more than one coalition. Moreover, these coalitions are dynamic and take place in real-time, after capturing customers' preferences. However, a coalition may succeed or fail. This primarily depends on the coalition's result, which can be complete or incomplete:

Definition 5: Complete vs. incomplete coalitions
We say that a coalition is complete if the coalition formed by the module agents builds up a salable product variant. A coalition is incomplete if the coalition formed by the module agents does not build up a salable product variant.

It is noteworthy that an agent will join a coalition only if the payoff it will receive from participating in the coalition is greater than, or at least equal to, what it can obtain by staying outside the coalition (Shehory/Kraus 1995, p. 58). In our case, a module agent that does not participate in any coalition has a payoff of zero. Because the module agent's account diminishes in the course of time, each agent should be interested in participating in beneficial coalitions in order to reconstruct its resources and thus to better ensure its existence. However, the success of coalition results is not certain. As aforementioned, each agent has to pay a certain fee for the participation in a coalition. But if the coalition that subsequently forms is incomplete or fails because it is not powerful enough to be a member of the final subset presented to customers, then the participation of an agent in a coalition is a waste of resources. Therefore, each agent has to be capable of estimating in advance the likelihood of the success of the coalitions it joins.

Recapitulating the assumption and definitions that are presented up to now, we can say that the upper goal of a module agent is to ensure a "long life" (Note the correspondence to variety steering!). This goal can only be reached if the module agent has enough resources for survival. These resources constantly diminish in the course of time. In order to reconstruct them, the module agents have to form successful coalitions. The coalition is successful if it is complete and also a member of the final subset to be presented to the customer (Note the correspondence to variety formation!).

2.2 The Main Framework for the Multi-agent Based Approach

In order to work efficiently, the multi-agent based approach has to be supported with a suitable technical framework (Figure 6-8). In addition to the module and platform agents, the framework consists of four other agents, namely the target costing, auction, product constraints' and validation agents, as well as two other components, namely the advisory component and configurator.

The advisory component is the starting point of the framework. It ensures the online supplier-customer communication by initiating a dialog with the customer in order to elicit his or her requirements. Then, the advisory component maps the customer's requirements onto product functionalities. Therefore, a software agent is needed to map functional requirements onto a technical product description.

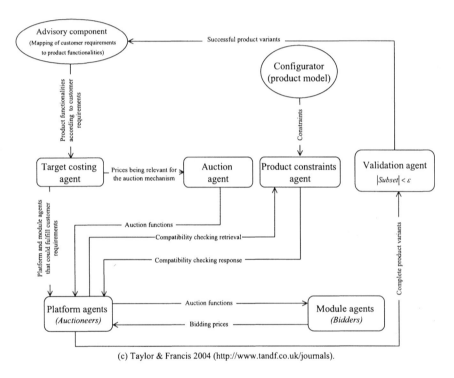

(c) Taylor & Francis 2004 (http://www.tandf.co.uk/journals).

Figure 6-8. Framework supporting the multi-agent approach for variety formation and steering

(Source: Blecker et al. 2004g)

In this context, target costing (Seidenschwarz 2001) provides many interesting insights. It can be basically defined as a cost management tool for reducing product costs over the entire product life cycle. It also provides a calculation method that aims at improving product design and planning. Based on the target selling prices that the customer would accept to pay and the target profits that the supplier would like to achieve, the target costs are estimated. To apply the method, the product subfunctions as well as their individual contributions in percent to the overall product function have to be determined. In addition, each product component has to be evaluated with respect to its contribution level to the fulfillment of each subfunction. Thus, starting from the product target cost, it is possible to estimate the allowed costs of each product component. Recapitulating, by considering target costing as a process, two main results are attained, namely a target cost of each component and a structured procedure that maps product functions onto technical product specifications. The adaptation of the target costing concept to our specific case provides a method that can automatically and dynamically translate the functional characteristics to product-oriented specifications. The software agent that executes this method is referred to as the target costing agent. It receives the product subfunctions from the advisory component and assigns to each of them the best module variants by considering the price range that the customer would be willing to pay. Thus, during an advisory session, the target costing agent determines the platform and module agents that are allowed to participate in the formation of coalitions. However, due to the modular architecture, the combination of the modules and platforms can generate too many product variants. In order to select good combinations, a refinement process is necessary. Therefore, the platform and module agents have to reach agreements concerning which coalitions to be formed. We propose to design the reaching agreement process according to an auction mechanism that coordinates the electronic market defined by the platform and module agents.

Auctions are "...a powerful tool that automated agents can use for allocating goods, tasks, and resources" (Wooldridge 2002, p. 131). They are also a successful means for coordinating electronic markets (Stroebel 2000, p. 39). In fact, within the context of agent technology, auctions can be viewed from two different perspectives. The mechanisms of an auction can be defined as a resource allocation problem to a set of agents or as a process of automatic negotiation implemented as a network of intelligent agents (Sandholm 1999, p. 84). The software agent that defines the auction mechanism is referred to as an auction agent. It receives the module variants' costs from the target costing agent and derives the auction functions that specify the protocol according to which the auctions run.

In the defined electronic market, we argue that platform agents are most suitable for initiating coalitions of module agents. Platforms are associated with high development times and costs and are considered to be the common basis of a product family for a long period of time. Therefore, it is legitimate to assume that platform agents dispose of an infinite account of monetary units and that, in contrast to module agents, they do not have to care about their existence. In effect, the introduction or elimination of product platforms is a strategic decision that has to be exclusively made by human managers.

The platform agents are considered to be the auctioneers in the defined electronic market, whereas the module agents are the bidders. For the initiation of coalitions, the platform agents have to be informed by the target costing agent about the module agents that are allowed to bid. Furthermore, they need to receive the auction functions from the auction agent. It is noteworthy that the type and number of product platforms that are allowed to initiate coalitions are ascertained by the target costing agent by considering the following proprieties:

- a platform agent can be selected more than once,
- each product variant is based on one platform,
- each platform can be found in several product variants and,
- the total number of selected platform agents is the utmost limit of the product variants that can be formed during coalitions.

The module agents bid in order to participate in the formation of coalitions. The platform agents as auctioneers also regulate the bidding process. When a module agent wins the bid, the platform agent asks the product constraints agent to verify if the corresponding module variant violates the constraints. If there are no compatibility problems between the module variants, the module agent is then allowed to join. Otherwise, the module agent is rejected and the auction will continue normally until a module agent with no compatibility problems wins the bid. For ensuring that only consistent product variants form, the product constraints agent must have direct access to the configurator's logic which contains the constraints between the different components in order to allow only consistent and completely structured product variants.

After all of the bidding processes terminate, some product variants may be incomplete because relevant modules are missing. These product variants are excluded and only the complete ones are communicated to the validation agent which selects the suitable subset of product alternatives. Then, the advisory component displays to customers the selected product variants.

2.3 Description of the Coordination Mechanism

For the description of the coordination mechanism, we will separately deal with the main agents that are involved in the process of reaching an agreement, namely the auction agent, platform and module agents.

2.3.1   The Auction Agent

In this section, we identify the main capabilities and characteristics that the auction agent should have. In the technical literature of auctions, one can make the distinction between two principle auction mechanisms, which are the ascending-bid and descending-bid auctions. In ascending-bid auctions, the bidders raise the price of the object by providing higher and higher bids, until only a single bidder remains. The price paid is equal to the last bid. In contrast, in descending-bid auctions, the auctioneer starts with a high initial price and progressively lowers it. The first bidder is the winner and the object is taken at the prevailing price (Klemperer 1999, pp. 229). The most common auctions that are successfully implemented in real world transactions are: the English auction, Dutch auction, first-price sealed-bid auction and Vickrey auction. In order to determine which auction mechanism is suitable for reaching an agreement between module and platform agents, we first identify the main characteristics to be fulfilled by the auction:

- The auction mechanism should enable a progressive formation of product variants. This enables platform agents to check compatibility while product variants form.
- The auction should enable one to ascertain in advance the auction length of time so that it is possible to adjust auctions in such a way that they terminate at the same time. It is relevant that product variants simultaneously form because the successful ones should be displayed at once to the customer.
- The auction should enable module agents to track the product variants while forming. In this way, module agents are able to better evaluate their chances of success.
- The auction should drive the module agents to bid as early as possible. Because of the product constraints, a module variant that bids early tends to better ensure its participation and avoids the restrictions that could be imposed by other module agents.

With respect to the first criterion, the first-price sealed bid and Vickrey auction are not suitable auction mechanisms. These cannot enable product variants to form progressively because the information of bidders is not open. Furthermore, the disadvantage of the English auction is that it does not permit a prior estimation of how long the auctions will last. However, the Dutch auction mechanism fulfills all of the proposed requirements. It is an open auction where product variants can progressively form. The auction's length of time can be well adjusted by fixing the starting price and the decreasing money amount in the course of time. In addition, the Dutch auction drives agents to bid as early as possible in order to increase their chances of winning a bid. The corresponding auction agent in the proposed framework is called a Dutch auction agent. The Dutch auction agent defines the functions with which the platform agent initiates the bidding process. Platform agents communicate these functions to all module agents, which are allowed to join the auction. For each auction, the Dutch auction agent generates a corresponding, suitable function in the form of:

$$K_i(t) = K_i g_i(t)$$

where

$K_i(t)$: the Dutch auction function decreasing in the course of time,

K_i : a constant representing the first price of the Dutch auction,

$g_i : [0, BT] \rightarrow]0,1] / g_i(0) = 1$; a steadily decreasing function and,

BT : the maximal bidding time

The Dutch auction functions must have the following properties: (a) the start value is derived from the value of the product function that is provided by the target costing agent, (b) the initial price must be harmonized with the module agents' starting accounts, and (c) all auctions that run for the same customer session must end at approximately the same time. (a) implicates that more valuable product functions are more expensive for module agents. This is legitimate because the possible rewards are consequently higher. (b) demands that the Dutch auction agent should have the capability of mapping the monetary value provided by the target costing agent to the units used in the auction process. Therefore, the Dutch auction agent also plays an important interfacing role between all agents and the advisory component (i.e. the user). (c) is required because the auction process carries out product variant formation in real-time. The customer concurrently obtains product suggestions. This demands that all auctions terminate at the same time in order to avoid delays.

2.3.2   Platform Agents

As aforementioned, due to their infinite account, platform agents do not have to be concerned about their existence. Platforms are created to be the basic module of a wide range of product variants for a long period of time. For instance, the motherboard can be considered as the platform of a computer. It contains the slots in which the necessary components such as a processor, hard disk, and cache memory are assembled. Different computing powers can be generated on the basis of the same motherboard by simply interchanging different components. The motherboard can be used for a long period of time since the computer's performance mainly depends on the performance of the other components. In the automotive industry, platforms (e.g. A-platform of Volkswagen) are especially cost- and time-intensive. An automobile platform is used in a wide range of car models and variants (Bock/Rosenberg 2003, p. 2) and can include, for example, the chassis, transmission, etc.

The elimination of a product platform would generate the elimination of all corresponding models and variants that are based on this platform. This decision is strategic and should not be allocated to software agents. However, it is conceivable to design the platform agents in such a way that they strive to be successful as much as possible, e.g. by contributing to the most sales' volumes.

Platform agents decide if a module agent is allowed to participate in a specific coalition or not. Therefore, they are more powerful than module agents. Platform agents initiate and coordinate the formation of coalitions based on information that is provided by the target costing and Dutch auction agents. They should also be capable of communicating with each other in order to avoid the formation of identical coalitions (product variants). Platform agents have the overview of the coalitions while forming and can forbid the further bidding of module agents by taking into account the constraints imposed by module agents that have already joined the coalition.

Recapitulating, platform agents are essentially concerned with the coordination task during the variety formation process. They steer the reaching agreement process with the module agents and communicate with other agents in the technical framework, namely the target costing agent, Dutch auction agent and product constraint agent. Consequently, platform agents mainly perform communication tasks.

2.3.3   Module Agents

In opposition to platform agents, module agents should maximize their utilities (accounts) by developing the appropriate strategies in order to ensure their survival. They should also be able to evaluate in advance the success of the coalitions by estimating the probability that the formed product variants take part in the final subset. Module agents have to ascertain which coalition would be beneficial to join as well as the optimal bidding point in time at a coalition. In order to formulate its bidding strategy, the module agent has to consider its account surplus. The account is private information of the module agent just as it is in real-world auctions. As a consequence, each module agent does not dispose of complete information. However, each agent should have the capability of tracking its environment in order to anticipate the behavior and the actions of opposing agents. Module agents have to update their knowledge and beliefs on the basis of their own experience, observed behavior of the other module agents and their current accounts.

A reward mechanism enables the module agents to compensate their continuously decreasing accounts. By winning a bid, the module agent pays the auction price that is considered as a fee of participation. At the end of all auctions, several product variants form. The number of product variants that come through the validation agent is inferior or equal to the number of all formed variants. Thus, the collected sum of monetary units from all bidding agents will be attributed to the few remaining variants. Subsequently, the amounts are divided on the module agents of the successful product variants. The reward that the module agent receives is equal to the difference between what it received and what it originally paid as a fee of participation. Consequently, some agents draw a profit, whereas some others incur a loss. The account of the unsuccessful agents diminishes more rapidly than those with positive rewards. Curve (b) of Figure 6-9 illustrates the progression of the account of a module agent that has participated at only successful product variants, whereas curve (c) provides the example of a module agent that has always been unsuccessful. Note that the life of the module agent having the account represented by Figure (b) is longer than the module agents having the accounts represented by (a) and (c). In order to determine the capabilities of the module agents, two types of strategies can be defined: a long-term and a short-term strategy. Whereas the long-term strategy has to ensure the agents' survival, the short-term strategy influences the agents' bidding behavior in running auctions. In the following, we deal in detail with both strategies.

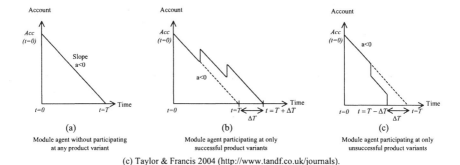

(c) Taylor & Francis 2004 (http://www.tandf.co.uk/journals).

Figure 6-9. Influence of product variants' successes and failures on the agent's account
(Source: Blecker et al. 2004g)

Long-term strategy

The long-term strategy refers to the plan that the module agent sets in order to achieve its fundamental objective that is to ensure a lengthy existence. To define the frames of the long-term strategy, variety managers have to ascertain two main entries for a module agent, which are *Acc (t=0)* and T_∞ (Figure 6-10). *Acc(t=0)* is the starting account that the module agent receives at the beginning of its life cycle. It determines the period of time T that the module agent would survive even when it participates in no auctions. However, T_∞ represents the period of time that the agent strives to survive. Within the scope of variety steering decisions, T_∞ is a relevant parameter that should be appropriately and carefully ascertained by variety managers. Owing to the rationality principal, the module agent strives to position itself on or above the fictive curve (2) of Figure 6-10. This way, the agent can ensure its survival until T_∞. This period of time can be attained when the agent achieves the net amount of monetary units: Pr*ofits* during its life cycle. The term $\dfrac{\text{Pr}\mathit{ofits}}{T_\infty}$ indicates how much the module agent has to attain per unit of time in order to achieve T_∞. The long-term strategies of the agents in a module class strongly depend on the value of this term. By assuming that the risk of failing is equally distributed among all agents in one class, the agent with higher $\dfrac{\text{Pr}\mathit{ofits}}{T_\infty}$ will tend to participate in more auctions and is more aggressive than the others. The aggressiveness level (Benameur et al. 2002, pp. 295) of a module agent characterizes its intention to bid rapidly in order to participate at product variants. Furthermore, the richer a module agent is, the more it tends to be aggressive. Thus, the aggressiveness level of a module agent depends both on its account at a time t, as well as the profits it has still to gain in order to achieve T_∞.

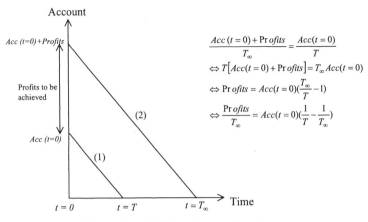

Figure 6-10. Module agent's long-term strategy

(Source: Blecker et al. 2004g)

Short-term strategy

In order to connect the long-term and short-term strategies of a module agent, we introduce the notion of a module agent's budget. A budget (*Bg*) is the sum of money that the agent allocates to bid in the currently running auctions. The allocation of budgets depends on the survival strategy of the module agent that tries to allocate budgets in such a way that it can reach T_∞. The following example illustrates the module agent's short-time strategy. Suppose that there are *n* auctions that begin at the same time *t=0* and in which the module agent would like to participate. The Dutch auctions start at a given price with a decreasing cost function. According to its strategy and aggressiveness level, the module agent ascertains a budget for these auctions and estimates the points in time when it bids. At *t=0* the module agent sets a plan and ranks the bidding times in an ascending way. When the auctions run, it may be that the module agent loses the first bid it has planned, or it is no longer allowed to participate because of product constraints. Thus, the module agent can reallocate the available budget differently for the participation in the remaining auctions. In order to illustrate the idea, we consider figure 6-11.

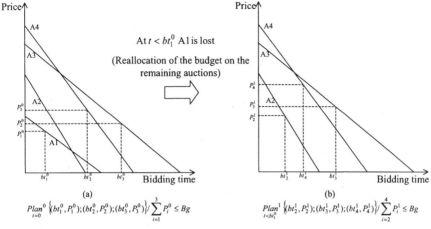

Figure 6-11. Module agent's short-term strategy

(Source: Blecker et al. 2004g)

Figure 6-11 shows that there are four auctions (A1, A2, A3 and A4) in which the module agent would like to participate. At *t=0*, the module agent develops a plan detailing where, when to bid and at which price. The module agent decides to participate in all auctions except A2. But at a point in time preceding the planned bidding time for A1, another agent wins the bid and/or imposes product constraints. Thus, the agent has the possibility to reallocate the released amount of monetary units. For example, this may enable it to participate in auction A2 and to increase the bidding prices for the other auctions provided that they are still running. It is important that the module agent learns from its failures in order to improve its future plans and budgets' allocation. By taking again the example of Figure 6-10, in the case when auction A1 would lead to a product variant success, whereas A2, A3 and A4 to failures, the budget reallocation would not be beneficial because it has not made the module agent more successful.

In the following, we formally illustrate the idea of short-term plans of the module agents. Therefore, we draw on a representation from decision theory that defines a rational agent as one that maximizes expected utility (Russel/Norvig 1995, p. 475). The expected utility *EU* of an agent's action is defined as:

$$EU(\alpha) = \sum_{\omega \in \Omega} U(\omega) P(\omega | \alpha)$$

where

$\alpha \in Ac$; Ac : the set of all possible actions available to an agent,

$\Omega = \{\omega, \omega', \ldots\}$: the set of all possible outcomes

$U : \Omega \rightarrow IR$ a utility function associating an outcome with a real

P : a probability distribution over possible outcomes given the performance of actions

Let the function f_{opt} take as input a set of possible actions, a set of outcomes, a probability distribution and a utility function and let this function return an action. The defined behavior of f_{opt} is defined as follows (Russel/Norvig 1995, p. 472):

$$f_{opt}(Ac, \Omega, P, U) = \arg\max_{\alpha \in Ac} \sum_{\omega \in \Omega} U(\omega) P(\omega | \alpha)$$

Wooldridge (2000, p. 10) criticizes f_{opt} for building rational agents because f_{opt} requires an unconstrained search that can be very expensive when the space of all actions and their outcomes is very wide. But, in our case this critique does not seem to be strong because the action types that a module agent can perform are (a) to participate in a coalition (Participating is the action $\alpha = 1$) or (b) not to participate (Not participating is the action $\alpha = 0$). Therefore, a module agent only has two types of actions. Furthermore, the outcome of actions may be either (a) the module agent is a member of a product variant which is selected in the final subset (Success of a coalition is the outcome $\omega = 1$) or (b) the module agent is a member of a product variant which is not selected in the subset to be presented to customers (Failure of a coalition is the outcome $\omega = 0$). That is why we argue that f_{opt} is suitable for the design of module agents. However, f_{opt} should be adapted to the requirements placed on the module agents.

Suppose at a point in time t=0 the platform agents initiate coalitions. Furthermore, suppose that the module agents that are allowed to bid have been already ascertained. To each module agent, the platform agent communicates the suitable auction functions $K_i(t)$ received from the Dutch auction agent. As aforementioned, when a module agent joins a coalition, it may restrain the participation of other module agents also intending to join the coalition. Therefore, each module agent must be capable of evaluating the behavior of the other agents that could prohibit its participation. In addition, each module agent has a certain aggressiveness level that depends on its current account and the profits it has to achieve in order to reach T_∞. Both aspects, namely the behavior of other agents in the environment and aggressiveness level are captured by the function *Risk* which is a risk function of a module agent MA_{ij} (Blecker et al. 2004b, p. 8):

$$Risk : [0, BT] \to [0,1] / Risk(0) = 0 \text{ and } Risk(BT) = R$$

where

Risk is a steadily increaing function

BT : Maximal bidding time

$R \in [0,1]$ is a constant reflecting the risk willingness of the module agent

A particular module agent is more risk-averse than another agent if R has a higher value and its *Risk* function more rapidly increases in the course of time (e.g. exponential). However, a module agent has a higher risk acceptance if R is relatively small and the *Risk* function slowly increases (e.g. logarithmic). The *Risk* function should drive the module agent to bid early in order to increase its chances of participation in a coalition. On the other hand, let Re *venue* be the function that takes the value 0 when the agent is not a member of the coalitions representing product variants displayed to customers and the value Re v when the product variants are displayed to customers. Thus, the utility function U of a module agent depends on the function $1 - Risk(t)$, which is supposed to decrease revenue during the auction process, revenue function and Dutch auction function. The adapted f_{opt} (Blecker et al. 2004b, p. 9) for our case is:

$$f_{opt}(Ac, \Omega, Risk, \mathrm{Re}\,venue, g_i, P, U) = \arg \max_{t \in [0, BT]} \sum_{\omega \in \{0,1\}} \{[1 - Risk(t)]\mathrm{Re}\,v - K_i g_i(t)\} P(\omega | \alpha)$$

The adapted f_{opt} returns the point in time t at which the module agent has to bid for the Dutch auction in order to maximize its utility. Note that if $t \in [0, BT]$, then $\alpha = 1$ and if there is no $t \in [0, BT]$ that maximizes the utility ($t > BT$), then $\alpha = 0$ and the module agent intends on not participating in the coalition. Furthermore, suppose that a module agent MA_{ij} is allowed to participate in p coalitions $C_k, k \in \{1, \ldots, p\}$. For each coalition the module agent estimates f_{opt}. At the point in time $t=0$ where an auction begins, the module agent develops a plan $(Plan_{t=0})_{ij} = ((\alpha_1^0, t_1^0), \ldots, (\alpha_k^0, t_k^0), \ldots, (\alpha_p^0, t_p^0))_{ij}$ which indicates whether or not and when to bid for each coalition (see figure 6-11).

For notation purposes, when $\alpha_k = 0$, the module agent allocates to t_k^0 an infinite value $\left(t_k^0 = \infty\right)$, which is in accordance with the fact that an agent will never bid and subsequently not to participate. By developing a plan the module agent has to consider the budget's constraint. This means that $\sum fees \leq Bg$. However, as stated previously, the agent's plan that is determined at $t = 0$ is not fixed for the entire auction process. The module agent adapts this plan according to the changes that occur in its environment. Suppose that the bidding times of a module agent MA_{ij} are arranged in an ascending way $(t_1^0 \leq ... \leq t_k^0 \leq ... \leq t_p^0)$ where $t_1^0 \neq \infty$ *and* $t_2^0 \neq \infty$. At a point in time $t < t_1^0$, a module agent from the same class may win the bid or an agent from another class may impose participation constraints. Subsequently, the module agent has to estimate once again f_{opt} for the remaining coalitions in order to determine whether and when to bid because when the participation in one coalition fails the module agent can reallocate the resources (Bg) that it has planed to expend differently. The resulting plan at a point in time $t = 1$ is: $(Plan_{t=1})_{ij} = \left(\left(\alpha_2^1, t_2^1\right), ..., \left(\alpha_k^1, t_k^1\right), ... \left(\alpha_p^1, t_p^1\right)\right)_{ij}$. This further illustrates the example depicted by figure 6-11.

2.3.4   Variety Formation and Steering Processes

An overview of the variety formation and steering processes is provided by figure 6-12. The multi-agent based variety formation process can be depicted by four main steps: (1) The advisory component, target costing agent and Dutch auction agent prepare relevant information for carrying out the auction mechanism. Note that the target costing agent does not generate information that exactly determines one single product variant. However, several variants may be possible. Depending on the sharpness of information that is gained from customers, the target costing agent can specify a class of module agents, a specific set of module agents in a class or even one particular agent in a class. (2) Then, the platform agents initiate Dutch auctions in which module agents bid in order to form complete coalitions representing product variants that could fulfill a customer's requirements. (3) These product variants are submitted to the validation agent which (4) selects the best ones to be displayed to customers by means of the advisory component. The validation agent has to base its selection task on data that is different from that of which the module agents have access to. By choosing the best subset, it reposes on the data that is gained online from customers through the advisory component, as well as on aggregated data that is retrieved from sources such as Customer Relationship Management (CRM) or Online Analytical Processing (OLAP) databases. However, the module agents develop their short-term strategies on the basis of information that is open during the run of auctions, as well as self-deductions with respect to the behavior of other module agents.

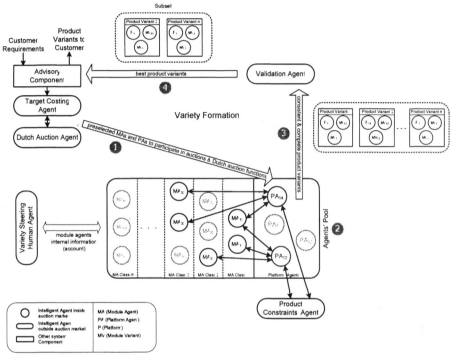

Figure 6-12. Overview of the variety formation and steering processes

(Source: Blecker et al. 2004f, p. 32)

The variety steering process considerably depends on the module agent's account. In effect, if the agent's account is exhausted, then the corresponding module variant should be selected for eventual elimination. The module variant should not be eliminated automatically, but human managers have to carefully examine this decision. It is noteworthy that the participation of a module agent in only a few coalitions does not enable one to draw good conclusions about the agent's success or failure. Therefore, the point in time, by which an agent shows a tendency to success or failure can only be estimated after simulation of the multi-agent system.

Furthermore, variety steering depends on two important values ($Acc(t=0)$ and T_∞) that must be carefully ascertained by variety managers. Both values consider decisions from business administration, which are relevant for the organization of the coordination process between platform and module agents. *Acc(t=0)* determines how long a module variant is allowed to exist in the supplier's offer even when no customer is interested. However, T_∞ is the longest period of time, during which the module variant is estimated to contribute to the fulfillment of customers' requirements. Because of e.g. new technological innovations or changing customers' preferences it is expected that after this period the corresponding module variants should be excluded from the offer.

3.   SUMMARY

The main problem fields that can endanger the success of the web-enabled mass customization are the external and internal complexities. External complexity is perceived by customers in extensive product environments, whereas internal complexity is experienced inside operations and manufacturing-related tasks. The external complexity problem is solved within the scope of variety formation, whereas the internal complexity problem is dealt with by variety steering. In order to cope with both problems at the same time, a decentralized information system supported by agent technology was recognized to be a suitable solution approach.

It was assumed that products are manufactured on the basis of modular and platform strategies. To each module variant and platform, autonomous rational agents are assigned. These agents are called respectively module and platform agents. In addition to these agents, the information system framework that is required for variety formation and steering includes a configurator, advisory system, target costing agent, Dutch auction agent, product constraints' agent and validation agent. All of the defined components and agents have to communicate and interchange data with each other.

The coordination mechanism was described through the specification of the properties and characteristics of the main software systems in the framework, namely the Dutch auction, platform, and module agents. The platform agents as auctioneers initiate the different auctions on the basis of the Dutch auction functions, whereas the module agents as bidders bid to take part in complete coalitions representing product variants that suit a customer's requirements. The module agents strive to ensure their survival for a long period of time. Only the fittest agents can ensure a lengthy exis-tence in the mass customizer's offer. Each module agent has to ascertain two types of strategies, which are the long- and short-term strategies. The long-term strategy is essentially determined by the agent's account, whereas the short-term strategy depends on the agent's budget. The agent's account contains a fictive amount of monetary units that continuously decreases in the course of time. In order to compensate the diminishing account, the module agents have to bid for the Dutch auctions to be able to gain a place in the forming product variants. However, only the best product variants that are able to come through the validation agent are considered to be successful. A reward mechanism recompenses the module agents that participate at the successful product variants. On the other hand, the agent's account provides managers with useful information about the success or failure of the module agents. On this basis, it is possible to draw conclusions about which variants should be retained and which ones have to be excluded (variety steering).

Due to the complexity of the mentioned problems in mass customization, the developed system is also complex. However, the theoretical system design is robust and feasible. For a successful implementation, it is necessary to accurately determine the initial values of the different parameters that are defined to make the system work. Therefore, some simulations are required before setting up these values. Furthermore, the system is to a great extent customer-oriented and the sharpness of the information would influence the agents' behavior. For instance, when the advisory component detects that the customer is a product expert, then the formulation of dialogs can be more product oriented and therefore the mapping of product functionalities to modules, which is carried out by the target costing agent is easier. In this case the costs' estimation of modules is accurate. However, when the customer is not a product expert, the sharpness of the gained information is lower. Nevertheless, indirect questions during dialogs about e.g. customer income enable the system to capture the cost range of the product variants that the customer would be interested in. Then, the information system refines the variants from the mass customizer's solution space to display only the product variants that would suit customers' requirements. Recapitulating, we can say that the main advantages of the developed multi-agent approach are:

- *Easy maintenance of the system*: By introducing or eliminating module variants, it is sufficient to introduce or eliminate module agents.
- *Dynamic variety generation during the interaction process.*
- *The application of a market mechanism:* This concept lets the intelligent agents themselves decide according to the current situation about their suitability to fulfill real customers' requirements. Such an approach represents an efficient coordination mechanism, even for a high number of involved agents (Shehory et al 1997, p. 144).

In the next chapter, we discuss different scenarios, in which the implementations of the stand-alone advisory system or multi-agent based system make sense. Furthermore, the integration possibilities of each system in the existing IT landscape of the supplier are described.

Chapter 7

IMPLEMENTATION SCENARIOS OF THE INFORMATION SYSTEMS

An efficient variety and complexity management has been identified as a necessary condition for the success of mass customization. By means of the information systems (advisory system and multi-agent system) that were developed in chapters five and six, the most relevant complexity problems were addressed, namely the external and internal complexities. The external complexity is perceived by customers, whereas the internal complexity is experienced inside the company's operations and manufacturing related tasks. The advisory system only addresses the external complexity problem by assisting customers in their buying decision. However, the multi-agent based concept copes with both complexity problems at the same time by forming suitable product configurations that meet specific customer requirements as well as supporting the mass customizer in making optimal decisions with respect to the variety steering task. In this chapter, we refer to the multi-agent *based* system as the system that includes all of the components that are necessary for variety formation and steering, excluding the configuration system, product constraints agent and advisory system. These components are: the target costing agent, Dutch auction agent, validation agent, module agents and platform agents.

The proposed information systems cannot be efficiently implemented without taking into account the supplier's specific situation and requirements. They have to be scaled and integrated into the supplier's existing information system landscape. In the context of integrated information management, the developed systems should support the information system infrastructure that already exists. At the same time, they need to be supported by the existing information system infrastructure. Therefore, the characteristics of the existing system have to be considered while making the decision about which system to be implemented. In this chapter, we depict two possible scenarios. The first scenario concerns the implementation of the advisory system and the second scenario deals with the implementation of the multi-agent approach. We also outline the advantages and disadvantages of each implementation case if certain preconditions are satisfied.

1. SCENARIOS AND CONDITIONS FOR A SUCCESSFUL IMPLEMENTATION OF THE INFORMATION SYSTEMS

Before depicting the possible scenarios according to which the proposed information systems can be implemented, it is first necessary to determine in which case the advisory system may be more suitable than the multi-agent approach and vice versa. We refer to the multi-agent *based* system as the system that consists of the target costing agent, Dutch auction agent, validation agent and agent pool which includes module and platform agents. Whereas the advisory system can be implemented as a stand-alone technical solution, the multi-agent based approach requires that an advisory component to have been implemented already. In effect, the mechanism designating the platform and module agents that are allowed to participate in the formation of product alternatives as well as the coordination process to a great extent depend on customer requirements. Therefore, an advisory component is necessary in order to make the multi-agent based system work efficiently. It is worth noting that the configurator, product constraints agent and advisory component are not considered to be elements of the multi-agent based system because of the following reasons:

- It is assumed that a product configuration system already exists. Furthermore, a configurator is a well-established software system that can be used in a stand-alone approach.
- From a technical perspective, it is more advantageous to separate the multi-agent based system from the configurator. In this context, the product constraints' agent should be seen as an interface between both systems.
- The advisory component can be implemented in a stand-alone version. It is conceivable to firstly implement the advisory system in a stand-alone approach and then to extend it with the multi-agent based system.

The implementation of a stand-alone advisory system or both a multi-agent based system and advisory system each has specific advantages and disadvantages. The stand-alone advisory system only deals with the external complexity problem, while the multi-agent based approach simultaneously copes with both of the complexity problems. Therefore, it can be stated that the multi-agent based concept is more powerful than the stand-alone advisory approach. As a result, the efforts and incurred costs required for the implementation of the multi-agent approach are higher. Furthermore, the decision about which system to be selected should basically depend on the specific requirements of the mass customizer. For instance, if the efforts and costs have to be maintained at a lower level and the main goal is to only provide customers with consulting services, it is more advantageous to opt for the less powerful, yet easier to be implemented system, namely the advisory system. The following criteria can be drawn on in order to support the decision concerning the information system to be selected:

- The specific requirements of the mass customizer in essence means whether the system should support both variety formation and steering tasks or not. This mainly depends on the complexity level within the company's operations.
- The existing information system infrastructure, which determines the available data that can be retrieved by the system to be implemented.
- The integration efforts that basically depend on the available technical interfaces.

The proposed multi-agent and advisory approaches cannot be considered as out-of-box solutions. Both systems have to be adapted to the IT infrastructure and integrated into the existing information systems. The complexity level of the IT landscape to a great extent determines the efforts that are necessary for integration. Therefore, the implementation must be rather regarded as a software engineering task.

In the following, we discuss the different scenarios, in which the implementation of the developed systems is adequate. Depending on the internal and external complexity levels, which can be either low or high, we identify three main scenarios as depicted by figure 7-1. If the external and internal complexities are low (Scenario 0), then product-oriented configuration systems are sufficient solution approaches. This case corresponds to a few product variants, from which customers are able to select optimal product configurations. However, low external and high internal complexity levels can correspond to two particular situations. In the first situation, the number of module variants is small and so all of the product combinations are straightforward for customers. The internal complexity is mainly due to the modules complexity, but not to high product variety. The second situation corresponds to the case, in which the number of product combinations is high, which involves high internal variety-induced complexity. However, the external complexity may be low if customers are perfectly aware of their requirements and can express them in terms of technical product language in spite of high product variety. In this context, it is important to note that customer expertise does not usually involve low external complexity. There are some cases, in which the configuration task is so complex that product experts have difficulties to determine the product configuration that exactly meets the requirements (e.g. an aircraft which is a very complex product with thousands of parameters to be fixed). In this case, external complexity cannot be considered to be low and customers should be supported with the appropriate information systems. Both situations that are depicted above correspond to the scenario 0 where the implementation of product oriented configuration systems is appropriate. In order to additionally cope with high internal complexity, the mass customizer has to use methods or tools such as the key metrics system that will be presented in chapter nine.

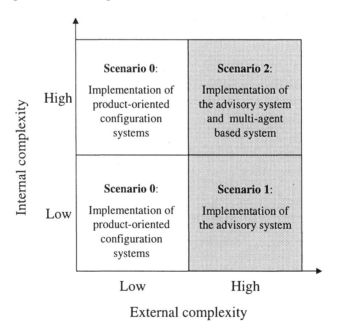

Figure 7-1. Appropriateness of the information system with respect to the levels of internal and external complexity

If the external complexity is high, we make the distinction between two scenarios. Scenario 1 corresponds to a high external complexity and low internal complexity levels. This refers to the case in which a high number of product variants can be built by only mixing and matching a few number of product modules. Thus, the mass customizer has to mitigate the external complexity problem by implementing advisory systems that assist customers during the decision making process. However, when the internal and external complexity levels are high (scenario 2), then an integration of the advisory component and multi-agent based system is a suitable solution approach. In the following sections, we concentrate on scenario 1 and 2, and especially on the implementation of the corresponding information systems that are identified to be suitable.

1.1   Scenario 1: High External Complexity/Low Internal Complexity

This scenario corresponds to the case, in which the external complexity is high and the internal complexity is low. A low level of internal complexity is for instance due to product variants that are built on the basis of a manageable number of modules. Therefore, an information system for assisting managers in the variety steering task is basically not needed. Instead, some simple managerial tools such as ABC analysis can be successfully applied in order to efficiently steer variety. In this particular scenario, an advisory system will completely suffice by supporting customers during the buying process. Furthermore, the advisory system is simpler than the multi-agent based system, which makes its management and integration easier and less costly.

The mass customizer is supposed to have already implemented an online configuration system. The mentioned drawbacks of configuration systems should be overcome when the customers are better supported through advisory. It follows that the advisory system has to be interfaced with the configurator. Although in the technical literature, one generally refers to the configuration system as a software tool that consists of both the product model and customer interface, we argue that both systems should be kept separated. The configurator contains the product model that only allows consistent product variants, whereas the advisory system, which is considered to be an independent software system, takes over the consulting role.

It is noteworthy that the configurators that are implemented up to now are equipped with customer interfaces in order to make the product model accessible to customers. For example, during the configuration of a product by means of a configurator that works with constraint-based reasoning, it can be noticed that after selecting a product option, the configurator

automatically restrains the number of remaining options due to predefined
constraints. In fact, this procedure reflects the internal logic of the
configurator. However, by means of an interface between the advisory
system and the configurator, both systems can be decoupled and the method
according to which the knowledge base of the configuration system is
modeled, does not affect the advisory process. Thus, the choice of the
product modeling method is a decision that can be made without considering
the customer's perspective. This is in accordance with the concept developed
by Leckner et al. (2004, p. 201) who argue that the Web-based configurator
tool should not be restricted to a specific product model. Instead, the tool
should dynamically adapt to an underlying product model. For instance,
when the product is modular, a resource-based configurator is advantageous.
But this representation may be not optimal for customers with a little or no
product-specific knowledge. The implementation of an advisory system
reduces the external complexity because it decouples the customer interface
from the product logic. The decoupling interface provides an abstraction
layer between the configurator and the advisory system and ensures that both
systems work as a unit that constitutes the whole interaction system.
However, it can happen that the suggested product alternatives do not
exactly correspond to the needs of the customers who prefer to introduce
some changes to the proposed product variants. The interface should also
consider this option that enables the customers to have direct access to the
configurator in order to change the product alternatives in terms of technical
specifications. Schematically, this can be depicted as follows:

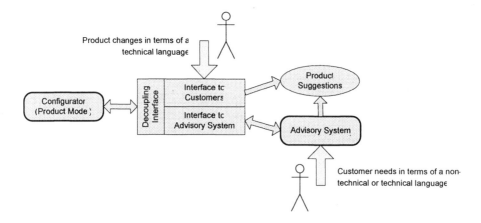

Figure 7-2. Conceptual architecture of the interaction system coping with external complexity

The decoupling interface provides an independent layer between the advisory system and configuration system. It is useful to provide an implementation-independent, well-defined interface between the heterogeneous systems, e.g. supported by CORBA services. The following communication steps are necessary when interfacing the configuration system and the advisory system whose implementation has already been described in a stand-alone version:

1. Initially, the product attributes contained in the configurator's product model must be exported to the repository of the advisory system, where they must be additionally annotated in the advisory's knowledge acquisition component. This is also relevant for the advisory process that captures customers' preferences and maps them to product specifications.
2. The data collected during the advisory process provide the main input for configuration. The advisory system initiates configuration by transferring the product specifications to the configurator that accordingly creates valid product variants.
3. Finally, the configuration results are returned to the advisory system that presents the product suggestions to the user. In order to refine the product suggestions, the advisory system allows customers to re-execute step 2.

Communication step 1 is an asynchronous task, whereas steps 2 and 3 must be carried out in real-time during the customer interaction. This leads to the following implementation approach:

- Product attributes of the configurator's product model are exported into the advisory system via XML.
- The Java framework is suitable for the implementation of the real-time interface between the advisory system and the configuration system. Furthermore, this interface is realized as an add-on to the advisory component that calls for services provided by the configurator.

Figure 7-3 outlines the architecture of the interaction system by depicting the interfaces described above, as well as the advisory system and the configurator. It extends the implementation concept of the extended advisory system through the interface to customers, which is attached to the configurator that additionally enables customers to introduce changes to the product suggestions in terms of technical language.

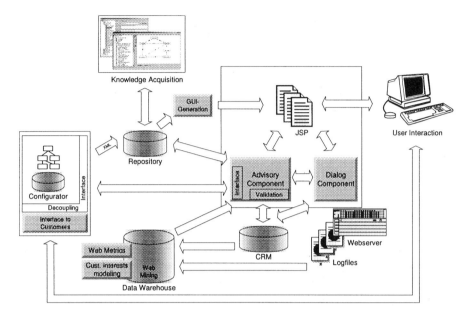

Figure 7-3. Technical architecture for an interaction system coping with high external complexity.

This technical architecture depicts the complexity of the proposed system with respect to the interfaces to be implemented. However, the use of standard interfaces such as XML has a great potential to shorten development times and costs.

1.2 Scenario 2: High External Complexity/High Internal Complexity

The only implementation of advisory systems does not enable the mass customizer to mitigate the internal complexity problem. However, it keeps the implementation efforts, time and the corresponding costs relatively low. Scenario 2 considers the case where both the internal and external complexity levels are high. The internal complexity problem endangers the success of mass customization. Therefore, it must be adequately addressed. The multi-agent based information system is capable of coping with both problems at the same time. The implementation is outlined by figure 7-4 which depicts the conceptual architecture of the information system by adapting and extending figure 7-2 of the previous section.

The system consists of the following independent components:

- The configurator that contains the product model.
- The decoupling interface that allows customers to carry out changes on the product suggestions in terms of a technical product description. Furthermore, the decoupling interface has a component (product constraints' agent) that ensures the communication between the multi-agent based system and the configurator, instead of the communication between the advisory system and the configurator, as compared with figure 7-2.
- The multi-agent based system that dynamically forms product variants during the advisory sessions and supports variety steering tasks in the course of time.
- An advisory system that is able to elicit the customers' needs and preferences according to the customer's product knowledge. It is worth noting that there are some differences between the advisory system that has already been described and the advisory system that has to be implemented within the multi-agent based system. In scenario 1, the advisory system elicits the customers' requirements and then determines the product alternatives that customers would be interested in, on the basis of predefined rules and/or the past buying behavior of customers. However, in scenario 2, the advisory system only elicits customer requirements, by asking customers the appropriate questions. The process of determining optimal and consistent product alternatives is no longer the task of the advisory system, but of the multi-agent based system.

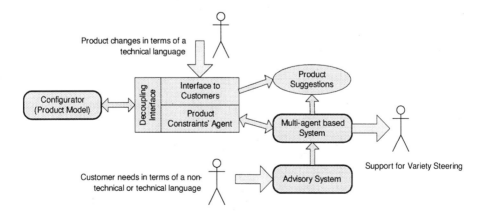

Figure 7-4. Conceptual architecture of the interaction system coping with external and internal complexity.

The product constraints' agent ensures the interfacing between the configuration system and the multi-agent based system. This interface intelligently acts in order to minimize communication overheads, as described in chapter six. The decision concerning the intelligence that is required to be built within this interface is mainly dependent on the implementation environment. For instance, if a high number of user accesses are expected, intelligent caching mechanisms are required in order to guarantee fast response cycles. Another alternative for interfacing both systems is to implement a web-service for checking product constraints. For an efficient work, the multi-agent based system requires information to be retrieved from several independent information sources. For instance, the information systems to be interfaced are:

- the CRM system which contains the customer profiles and properties and OLAP databases that aggregate customers' navigation behavior. Data that is retrieved from such systems is required by the validation agent for the selection of suitable product variants;
- the data warehouse that retains past transactional data that can be analyzed and retrieved by the target costing agent;
- the individually developed database applications that contain other relevant information.

The required data can be distributed over a multitude of heterogeneous and autonomous information systems. Therefore, ensuring a smooth data exchange is challenging. Hasselbring (2000, p. 35) recognizes that as a prerequisite, applications do not only need to understand the syntax of data provided by other applications, but the semantics as well. Thus, the standardization of message formats and message content plays an important role. In order to exchange data, XML has become the standard syntax on the Internet. Nevertheless, in order to provide more interoperability across applications, the semantics of the messages also have to be defined. Adam et al. (2000, pp. 68) identify three popular approaches for the integration of independent systems:

- **CORBA approach**: CORBA (Common Object Request Broker Architecture, www.omg.org) is a standard for application development within heterogeneous environments. It provides programming interfaces and packages that support heterogeneous platforms. In CORBA, services are abstractly defined and can be implemented platform-independently. At run-time, the CORBA framework redirects a client's request to the implementation of the service on a remote host. However, while requesting this service, the client does not need to know where and how the object is implemented. In the CORBA core, several standard methods for inter-process communication are supported, for instance shared memory, TCP/IP, or remote procedure calls. The development of CORBA-based application relies on the abstraction of interfaces using an Interface Definition Language (IDL).

- **Mediated approach**: In this approach, components called mediators perform the integration tasks. Therefore, so-called wrappers interact with information sources in order to convert queries from the mediator's language into the information system's native language. Thus, a wrapper provides an abstract interface to its underlying information source that can be exploited by the mediator. The mediator holds a common data model across all information systems and allows users to access this model. In order to perform an operation, the mediator component has to split up user queries into sub-queries and send them to the corresponding wrappers that transform them to the systems' native language. Then, the results are returned and combined into a complete answer by the mediator that is finally presented to the user.

- **Agent-based approach.** In order to be able to dynamically react to changes of the information sources, for instance by added or removed services, i.e. data sources, it is possible to follow an agent-based approach. Adam et al. (2000, p. 69) propose an architecture that extends the architecture developed by Wiederhold (1992) as follows: Mediator agents interact with interface agents and source agents. Source agents interface data sources that should be integrated, which means that they act as a kind of wrapper. Interface agents receive user queries and propagate them to a mediator agent. Subsequently, this agent contacts source agents in order to receive an appropriate response. To implement this approach, agent communication languages such as KQML, ontologies, or directory services can be considered. In opposition to the mediated approach, due to the self-organization of agents, this architecture is more dynamic in changing the underlying source systems.

Recapitulating, all of the approaches can be applied to the presented scenario. The CORBA, mediator and agent-based approaches each have specific advantages and disadvantages. The implementation based on CORBA is relatively simple. Furthermore, this technology is well known and widely supports the programmer. However, the main disadvantage is its inflexibility if the data sources change frequently. Since we can assume that the proposed architecture does not change after implementation for a certain period of time, the CORBA approach can be suitable. Nevertheless, the agent-based approach is more scalable and adaptable. Due to the fact that the variety formation and steering is already based on multi-agent technology, it is more appropriate to adopt this approach.

As aforementioned, the decision about the suitable technology to be implemented in order to integrate the information systems is primarily dependent on the available infrastructure. Therefore, the implementation problem should be tackled by taking into account the specific case and the existing IT landscape of the mass customizer.

2.   SUMMARY

In this chapter, we have discussed the scenarios, in which the implementation of the advisory system and the multi-agent based system for mass customization is appropriate. Whereas the advisory system is suitable when the external complexity is high and the internal complexity is low, the multi-agent based system is an adequate solution approach when both external and internal complexities are high. In comparison to the advisory system, the multi-agent based system has a higher complexity because it has to retrieve data from several autonomous data sources that must be interfaced. In addition, we have depicted the technical infrastructure that can be required in order to realize the integration of the advisory and multi-agent based systems in the existing IT landscape of the mass customizer.

PART III

CONCEPTS FOR IMPLEMENTING AN EFFICIENT PRODUCT CUSTOMIZATION

Chapter 8

PRODUCT MODULARITY IN MASS CUSTOMIZATION

Up to now we have only dealt with product modularity as the ability to configure product variants by mixing and matching components within a modular product architecture. Product modularity enables the manufacturing of a large number of product configurations by simultaneously taking the advantage of the economies of scale and scope. In fact, the concept of modularity has played a decisive role in the development of the mass customization paradigm. It enables not only the ability to put the "mass" in mass customization, but also to configure the products according to the customer's requirements. Whereas the advances realized in the field of information technology can be considered as an important enabler of mass customization, product modularity is considered to be a necessary requirement. Furthermore, modularity is not restricted to the product level by combining modules in order to create many variations. "Modularity refers to an ability to "decompose" technological and organizational systems such that the internal functioning of one subsystem does not significantly affect the functioning of the others in the short term" (Garud/Kumaraswamy 2003, p. 68). Thus, product modularity involves additional implications on the organization itself and on the way to conduct business. Due to all of the depicted reasons, it is important to deal in more detail with this concept and to discuss some relevant related aspects.

In this chapter, we will discuss the benefits as well as drawbacks of product modularity. Then we will deal with the relevant managerial implications of product modularity in mass customization. Finally, we will present some selected methods from the technical literature for a systematic development of modular product architectures. These methods deal with the main question of how to implement modularity into the design of product families with several product variations.

1.   PRODUCT MODULARITY

In the previous chapters of this book we have used a definition of product modules as building blocks that enable the configuration of a large number of product variations. This definition was sufficient and adequate for the development of the already presented ideas, especially those relating to the multi-agent system concept for variety formation and steering in mass customization. However, it is generally argued that there is no standard definition for the terms module, modular architecture, modularity and modularization (e.g. Ericsson/Erixon 1999, p. 19; Nilles 2001, p. 123).

In the following, we will provide definitions identified in the technical literature that are consistent with our understanding of the corresponding terms. But before that, it is relevant to define the term product architecture. It is "... the scheme by which the function of a product is allocated to physical components" (Ulrich 1995, p. 420). The main purpose of a product architecture is to define the product building blocks by specifying what they do and how they interface with each other. The choice of a product architecture plays an important role and to a great extent affects the performance of the manufacturing firm.

Nilles (2001, p. 127) points out that a product module is characterized by the following properties:

- A product module is a subsystem with lower complexity than the overall system of which the module is a part.
- A module is a closed functional unit.
- A module is a spatially closed unit.
- A module has well-defined and obvious interfaces.

Ulrich (1995, p. 422) mentions that a "...modular architecture includes a one-to-one mapping from functional elements in the function structure to the physical components of the product, and specifies the decoupled interfaces between components."

Ericsson/Erixon (1999, p. 19) define product modularity as "...having two characteristics: 1) similarity between the physical and functional architecture of the design, and 2) minimization of the degree of interaction between physical components." Baldwin/Clark (2003a, p. 149) include the process perspective in their definition of modularity. Modularity refers to "... building a complex product or process from smaller subsystems that can be designed independently yet function together as a whole" (Baldwin/Clark 2003a, p. 149). Both definitions of modularity are suitable. Nevertheless, we opt for the definition of Ericsson/Erixon (1999) when addressing product modularity because it explicitly includes a very relevant aspect of modularity that relates to the interfaces. The definition of Baldwin/Clark (2003a) is appropriate when dealing with the process according to which the modular product is developed.

However, modularization is the "...decomposition of a product into building blocks (modules) with specified interfaces, driven by company specific strategies" (Ericsson/Erixon 1999, p. 19).

By addressing the module strategy, it is necessary to define the mapping between the functional and physical elements of the product as well as the interfaces between modules in the product. Ulrich (1995, p. 424) distinguishes between three types of modular architectures, namely slot, bus and sectional, which involve different types of interfaces.

In a slot architecture, each of the interfaces between the components is of a different type from the others, which makes it impossible to interchange the various components of the product (e.g. the radio module in a car). The bus architecture is characterized by a common bus to which physical components are connected via the same type of interface (e.g. personal computer). In opposition to the bus architecture, the sectional architecture does not involve a single module to which the components can attach. The assembly is realized by connecting the components to each other via identical interfaces (Ulrich 1995, p. 424). Figure 8-1 illustrates the different types of architectures as well as the involved interfaces.

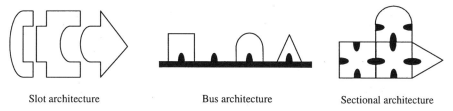

Slot architecture   Bus architecture   Sectional architecture

(c) Elsevier 2002, Reproduced with permission.

Figure 8-1. Modular architecture types
(Adapted from: Salvador et al. 2002a, p. 552)

On the basis of a case studies' analysis, Salvador et al. (2002a, pp. 558) find that slot modularity as identified by Ulrich (1995) is a general case that can be further detailed by defining a spectrum with two extremes, namely component swapping modularity (Ulrich/Tung 1991) and combinatorial modularity (Salvador et al. 2002a). Component swapping modularity refers to an architecture with a basic product body that is common for all product variations and one single family of several component variants with identical interfaces to the common part of the product. However, a product architecture that is based on combinatorial modularity has no common product body. The different product configurations are obtained by mixing and matching component (or module) variants whose interfaces are standardized by parings of component families. This means that the interface between two component families only depend on the particular coupling between the families, but in no way on the component variants that are selected to be combined. Combinatorial modularity is an ideal case that cannot be easily achieved in practice because of technological constraints that may hinder the interface standardization across all of the possible pairings of coupling the module family variants. Whereas the component swapping modularity has a high commonality level between product variations due to the basic product body, commonality in product architectures that are based on combinatorial modularity has a lower level. Thus, the spectrum of product architectures in slot modularity (Figure 8-2) is characterized by decreasing commonality when moving from swapping modularity to combinatorial modularity (Salvador et al. 2002a, p. 561; Salvador et al. 2002b, pp. 63).

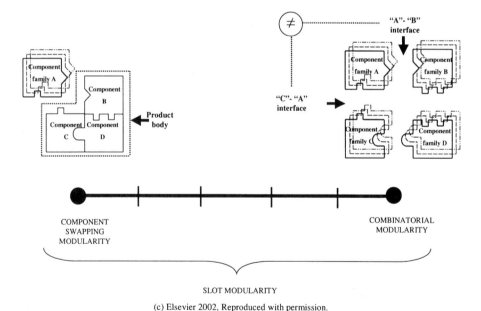

(c) Elsevier 2002, Reproduced with permission.

Figure 8-2. A graphical representation of the component swapping modularity-combinatorial modularity spectrum

(Source: Salvador et al. 2002a, p. 560)

Wildemann (2003, pp. 58) refers to the modules strategy as a relevant approach for an efficient management of the product arrangement system. In addition to the modules strategy, Wildemann also identifies five other strategies, which are the part families, common parts, building blocks, systems, and platforms. He proposes a classification of these approaches according to three main criteria, which are the potential for saving costs, flexibility and complexity reduction. The classification of Wildemann suggests that the module and system strategies have the same potential to reduce costs and complexity by ensuring the same flexibility level. However, in opposition to systems that consist of components that are not assembled together (e.g. a braking system including the pedal, booster, master cylinder, front and rear brakes), modules are spatially closed units (e.g. the engine of a car). Thus, a system may involve more interfaces than a module, which implies higher complexity and less potential to save costs. Therefore, we argue to separate both strategies in the classification proposed by Wildemann (2003) as it is shown by figure 8-3. All of the strategies proposed have not to be understood as a stand-alone approach. In effect, Wildemann (2003, p. 59) emphasizes that these strategies have to be combined in a coherent and efficient way in order to increase efficiency.

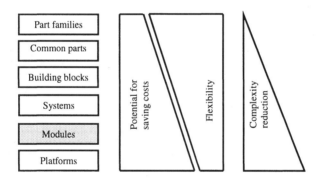

Figure 8-3. Classification of the modules strategy with respect to cost savings, flexibility and complexity reduction in comparison to other strategies

(Adapted from: Wildemann 2003, p. 59)

Product modularization is more often discussed in connection with the concurrent engineering concept. Whereas in traditional development processes, decisions are executed sequentially, in concurrent development processes, decisions are made simultaneously and knowledge from different domains is integrated. Concurrent engineering basically integrates product design and process design engineering at the early stages in order to develop products that can be manufactured easier at relatively lower costs.

Product modularity enables concurrent engineering, in that several development teams can simultaneously work on different product modules. This is due to the well-specified interfaces between the product modules. Each team can autonomously focus on one building block without having to continuously coordinate its tasks with other teams, which considerably decreases the development lead time as well as related costs (Hartley 1992).

2.   BENEFITS AND LIMITS OF MODULARITY

Product modularity has many advantages in mass customization. It enables one to put the "mass" in mass customization by combining a few modules on which basis a large number of product variants can be created. However, modularity has some drawbacks that managers have to take into account. In this section, we briefly discuss both faces of modularity as a powerful concept that must be handled with caution.

2.1   Benefits of Modularity

The benefits of modularity appear on both sides of the customer and producer. A product built around a modular architecture is open to innovation and involves the trying out of alternate approaches, which leads to rapid trial-and-error learning. Product modularity enables firms to be organized according to a decentralized network. More entry points are available for new firms with new ideas to join the network. Therefore, the entire performance of the product can be continuously and rapidly enhanced, especially by improving the bottleneck components that are critical for customer satisfaction (Langlois/Robertson 2003, p. 84). "...[I]n that network, local experiments can be used to try out new products and new production processes within the context of a (relatively) stable larger system. In this fashion, the network itself can evolve piecemeal, with most innovations causing only local disruptions in the overall system" (Baldwin/Clark 2003b, p. 38).

Modularity makes the realization of the economies of substitution possible by reducing such costs that would be incurred if components with no standard interfaces had to be matched with each other. Indeed, components with incompatible interfaces have to be adapted to each other by using gateway technologies, which involves additional costs. Economies of substitution arise when the costs associated with the design of a higher-performance system through the partial retention of existing components is lower than the cost of designing the system afresh. Components' retention, which can be attained through modular product architectures provides the possibility to reutilize the existing base of knowledge associated with these components. Other benefits are essentially due to savings in testing and production costs (Garud/Kumaraswamy 2003, p. 48).

Baldwin/Clark (2003a, p. 162) identify three main purposes that justify the incurrence of expenditures in modular architectures. First, modularity makes complexity manageable through the decomposition of complex products into smaller subsystems that can be independently designed. Second, it enables parallel work, in that several design teams can simultaneously work on the development of different modules. Third, modularity is tolerant of uncertainty, which means that particular elements of the modular design can be changed in unforeseen ways as long as the design rules are respected.

In the mass customization context, Piller (1998, pp. 194) depicts many advantages of product modularity. A modular architecture enables the postponement of product differentiation. Thereby, production lead times can be reduced because the different product modules can be concurrently manufactured. Product modularity contributes to the improvement of the quality of products as each module can be independently tested and possible defects can be discovered at the early stages. It is also an important marketing-tool because customers have the possibility to further individualize the product by simply replacing some modules through others. In the after-sales phase, the modular architecture improves maintenance and repair since defects or upgrades can be easily and rapidly carried out. Furthermore, product modularity facilitates the utilization of suppliers' know-how. Such a modular sourcing concept leads to decreasing fix costs and increasing flexibility, which involves some manufacturing capacities that can be released.

2.2 Limits of Modularity

Garud/Kumaraswamy (2003, p. 71) point out that modularity is an important attribute of technological systems. However, modularity is not the only system attribute that has to be taken into account. There are other attributes, which are also highly relevant, namely integrity and upgradeability. Integrity is "the consistency between a product's function and its structure: the parts fit smoothly, components match and work well together, the layout maximizes available space" (Clark/Fujimoto 1990, p. 108). The upgradeability that gains more in importance in rapidly changing environments is "...the ease with which system performance can be enhanced over time" (Garud/Kumaraswamy 2003, p. 47). It is worth noting that tradeoffs exist among all of the system attributes. For example, excessive modularization may trigger complex non-linear interactions among components, which considerably inhibits upgradeability. Thus, the modularity level embedded into a product design may have negative impacts on the other system attributes of high importance.

In the context of mass customization, Piller (1998, p. 197) notes that building products around modular architectures involves several potential dangers. In effect, complexity costs can be effectively reduced when a high number of individual products can be offered by using only a low number of modules. Furthermore, all customer needs cannot be fulfilled when developing products around modules because the variation occurs within certain combinations specified in advance. Another potential danger is due to the relative ease to imitate modular designs from competitors. Because of the high level of knowledge sharing when applying modularity, a potential risk is that firms become trapped within the confines of old knowledge. Thus the mass customizer continues to use only standard modules, which may have a negative impact on innovation. In addition, the development of modular systems is cost-intensive compared to integral systems. Garud/Kumaraswamy (2003, pp. 50) mention that modularly upgradeable technological designs alone are not adequate for the achievement of the economies of substitution. Modular designs incur additional costs consisting of the initial design, testing, and search costs. The initial design costs refer to the extra costs necessary for the design of carry-over components in comparison to components for a one-time use. Since testing costs are triggered at the component level and not at the system level, they increase with the number of modules involved in the modular architecture. Finally, search costs are incurred because designers have to search in component libraries for components that are candidates for reuse. However, it is relevant to mention that efficient and well-planned modular product architectures avoid these problems.

3.   MANAGERIAL IMPLICATIONS OF MODULARITY

Product modularity enables managers to reduce internal complexity, increase flexibility and create a proactive organization. The modularization of products decreases development lead times, which significantly improves time-to-market – the length of time needed to get a product from an idea to the marketplace. Product modularity provides an additional flexibility, in that some of the target values for the product properties that might not be fulfilled in time for the first product generation can be further developed after the introduction of the product to the market. Instead of delaying the launching date, the stepwise development for the technical solutions bearing this property should be outlined (Ericsson/Erixon 1999, p. 94).

The product architecture to a great extent influences the company's organization. Indeed, the design of every product necessitates specific tasks to be performed in order to develop and realize that product. Consequently, the product designs that a firm creates greatly influence the organization that the firm could adopt in order to develop, produce, distribute and service its products. In the case when the interfaces between components are complexly interdependent, then the organization should be designed in such a way that it supports frequent communication between the design groups. However, a modular product design implies standardized interfaces, which means that the design teams do not have to intensively communicate and can work independently. Due to this high level of autonomy, the design of the organization has to support the autonomy between the component development groups instead of frequent communication and coordination (Sanchez/Collins 2001, p. 660).

Product modularity enables the achievement of the economies of substitution. However, this goal can be attained only if the organization succeeds in devising and implementing adequate mechanisms. Garud/Kumaraswamy (2003, p. 46) note that managers have to elaborate incentives in order to encourage the reuse of already developed modules and components across many generations of product families. A key challenge in realizing economies of substitution is the design of organizational systems that enhance component retention or reuse while reducing the associated costs. At each product development project, managers have to define in detail the reuse targets in order to fully utilize the economies of substitution. If managers assess performance only through the degree of innovation, then design engineers will tend to develop new components even when existing ones are appropriate. However, if performance evaluation additionally takes into account the degree to which development engineers reutilize components, reuse will be improved, while achieving the economies of substitution. Therefore, managers have to institute effective mechanisms to search for reusable components, e.g. through component libraries.

A further relevant aspect of well-defined modular architectures is that they provide a powerful framework for identifying and leveraging a firm's current knowledge, for discovering hidden "capability bottlenecks" and for improving strategic learning processes. Indeed, the careful definition of the capabilities involved in creating modular product and process architectures makes the hidden "capability bottlenecks" that limit a firm's ability to design, produce and support new products rather apparent. The capability bottlenecks that are currently restraining a firm's options for the creation of new products can be targeted for focused strategic learning and capability development (Sanchez/Collins 2001, pp. 655).

Sanchez/Collins (2001, pp. 655) mention that management has a new role when products are developed around modular approaches. In traditional product development processes, the input of managers at the first steps of the design process is commonly low. However, their input considerably increases after the system design because of a lack of stable component interfaces, which triggers a chain reaction of component redesigns. In effect, the mid-level managers have to continuously make decisions about a variety of interface issues between development groups, which consumes significant amounts of time. By contrast, a modular architecture necessitates a high level of management inputs for the specification of the product platform, interface standards, required upgrading measures during the product life cycle, etc. Using the terms of Baldwin/Clark (2000, pp. 72; 2003a, p. 151) managers only have to intervene in the definition of the "visible design rules (also called visible information)" that are necessary to ensure the compatibility of the total design. After specifying the framework of the modular system, development teams can work independently and middle management does not need to intervene in order to solve problems that relate to interface issues. The developers only make decisions about "the hidden design parameters (also called hidden information)" as stated by Baldwin/Clark (2000, pp. 72; 2003a, p. 151) that do not affect the design beyond the local module.

Product modularization not only has an influence on an intra-firm level, but also on an inter-firm level with respect to the relationship with suppliers. On the basis of a case studies analysis, Salvador et al. (2002a, p. 572) examine the impacts of slot modularity on an inter-firm level in the case of an assembler with large production volumes that sources the constituent components of its products from external suppliers. They find that the nature of interaction between suppliers and a firm can be affected by the type of product modularity. The authors propose that "...[a]s the type of modularity embedded in the product family architecture moves away from component swapping modularity towards combinatorial modularity..., the extent to which the negative effect of variety on operational performances can be mitigated depends upon ... [t]he extent to which the component family supplier with the longest throughput time can be located in geographical proximity." (Salvador et al. 2002a, p. 572). Furthermore, they posit that in the case of component swapping modularity, the component family suppliers tend to be smaller or directly controlled by the final assembler. However, the firms that build their products around a combinatorial modularity tend to set up bilateral relationships with suppliers of the component families.

In this context, Baldwin/Clark (2003a, p. 153) mention that product modularity involves significant changes in the organization of the supply chain. They report the example of Mercedes-Benz during the planning of its new sport-utility assembly plant in Alabama. Mercedes' managers have realized that the complexity of the vehicle would require the control of a network of hundreds of suppliers by keeping high inventory levels because of demand uncertainty. Therefore, they have organized the supply chain in a set of large production modules where suppliers are completely responsible for the delivery of modules. In this way, the complexity of the suppliers' network is considerably reduced and the planning accuracy is improved. Another example provided by Baldwin/Clark (2003a) is that of Volkswagen's truck factory in Resende, Brazil. Volkswagen provides the factory where the suppliers build all of the required modules and assemble the trucks. Volkswagen establishes "... the architecture of the production process and the interfaces between cells, it sets the standards for quality that each supplier must meet, and it tests the modules and the trucks as they proceed from stage to stage" (Baldwin/Clark 2003a, p. 153).

From the perspective of information technology, note that modularity is an important property that has enabled us to introduce the multi-agent system for variety formation and steering described in part two of this book. Without this property it would be difficult, impossible even, to develop and propose a feasible implementation of the system. If a modular product design were not assumed, this would mean the association of an autonomous rational agent with each product variant. Due to the large number of product variants that are involved in mass customization, the multi-agent population would be so large that the generation of sound solutions in real-time is impossible.

The modularity level also has an influence on the product model to be chosen. It is relevant to recall that the core component of the configuration system is the knowledge base, which consists of the configuration database and the configuration logic. Thus, a product with a high modularity level involves a high number of standardized interfaces, which considerably decreases the complexity of the configuration logic that specifies the rules and the constraints according to which the components are connected to each other. Furthermore, the generation of the product variants' prices in real time for customers during the interaction process is also facilitated when products are based on modular architectures. It can be concluded that product modularity has important implications with respect to the software tools that are to be developed for mass customization.

4.   SELECTED TOOLS FOR THE IMPLEMENTATION OF PRODUCT MODULARITY

In the technical literature, it is argued that there is only a few tools and methods that deal with the question of how to partition systems in order to obtain modular product architectures (Dahmus et al. 2001, pp. 411; Ericsson/Erixon 1999, p. 27). In the following, we will describe two selected tools that represent structured methods for developing modular product designs. Both tools have been already implemented in practice with success and are based on qualitative rather than quantitative approaches.

4.1   Modular Function Deployment (Ericsson/Erixon 1999)

Modular Function Deployment (MFD) is a structured method that aims at identifying and optimizing product modules within the product assortment. It consists of five main steps as described by figure 8-4. The first step refers to the definition of customer requirements. At this stage, the product strategy and the brand image have to be defined. Ericsson/Erixon (1999, p. 32) report that a simplified version of the quality function development that translates customer requirements into product properties is an adequate approach at this step.

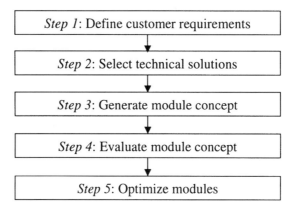

Figure 8-4. Modular Function Deployment procedure

(Ericsson/Erixon 1999, pp. 29)

The second step of the method deals with the selection of technical solutions. On the basis of customer requirements identified in step 1, the necessary product functions and sub-functions are derived. Then the corresponding technical solutions for each sub-function are selected. Thus, the result of this step is a functions-and-means tree that visualizes the functions-hierarchy of the product and the corresponding technical solutions. It is relevant to mention that within an ideal modular product architecture, there is a one-to-one mapping between product functions and modules. Therefore, the functions-hierarchy can be considered as an ideal modular decomposition of the product.

Step three is the core step of MFD. It aims at assessing the technical solutions selected for each sub-function (potential modules) with respect to the reasons of grouping product elements into modules, namely the module drivers. The module drivers are the driving forces for modularization within the product and cover the entire product life cycle. Figure 8-5 depicts the module drivers as well as a short explanation to each one.

The results of the evaluation of the technical solutions against the module drivers are represented in the Module Indication Matrix (MIM) (Ericsson/Erixon 1999, p. 34). Each technical solution is weighted on a scale where nine points are allocated to a strong driver, three points to a medium driver and one point to a weak driver. Ericsson/Erixon (1999, p. 35) use this scale in order to support the identification of strong module drivers. Highly weighted drivers would indicate that the technical solution is likely to form a module by itself. On the basis of the matrix, different module concepts are proposed.

Step four is targeted toward the evaluation of the module concepts that are generated in step three. At this step, the module interfaces that have a large amount of influence on the final product and the assortment flexibility have to be examined. In addition, several economic factors must be taken into account. However, Ericsson/Erixon (1999, p. 39) argue that traditional approaches of economic accounting and activity based costing are not suitable for the assessment of the advantages and effects of modular assortments. Therefore, they propose a rough estimation based on metrics and rules for the evaluation of the module concepts.

Upper criterion	Module drivers	Explanation
Product development and design	Carry over	Parts that will be carried over from an earlier product generation to a later one.
	Technology evolution	Parts that are likely to change because of changing customer demands or technology shift.
	Planned product changes	Product parts that are intended to be changed and developed by the company.
Variance	Different specification	Parts that are responsible for customization and product variation.
	Styling	Visible parts that are used to underline product identity.
Production	Common unit	Parts that are used for an entire product assortment or for a large number of products.
	Process and/or organization	Parts that require the same specific production process should be clustered together.
Quality	Separate testing	The possibility of separately testing each module before delivery to final assembly.
Purchase	Supplier available	Make-or-buy analysis for subsystems in the product.
After sales	Service and maintenance	Parts that are exposed to service and maintenance may be clustered together to form a service module.
	Upgrading	Parts that are upgraded by customers in the future.
	Recycling	Environmentally hostile or easily recyclable material can be kept separate in specific modules to simplify the disassembly of the product for recycling.

Table reprinted with permission of the Society of Manufacturing Engineers/Modular Management AB,
copyright 1999

Figure 8-5. The Module Drivers

(Adapted from: Ericsson/Erixon 1999, p. 21)

In the final step (step five), a detailed specification is written for each module, including e.g. technical information, target costs, etc. On this basis, an improvement of the modular concept can be achieved by focusing on each module separately. Thereby, the optimization of the modules is attained through the introduction of enhancements at the component design level by applying tools such as Design For Assembly and Manufacture (DFMA).

4.2 The Concept of Dahmus et al. (2001) for the Generation of Modular Product Architectures

Dahmus et al. (2001, pp. 409) develop a systematic methodology to architecting a product portfolio, with the objective of taking advantage of commonality through the reutilization of modules across a product family.

The authors argue that the method is applicable for the development of a completely new product portfolio as well as for the augmentation of an existing portfolio with additional product variants. An overview of the portfolio architecting process is described in figure 8-6.

The process begins with the determination of the underlying technologies that should be utilized in the product (e.g. for the development of a family of screwdrivers, it should be decided at the very early stages if the screwdrivers are electrically powered or air-compressed powered). Then, it is relevant to establish the limits of the product family whose different variants must share modules. For each product in the family, one or several conceptual designs have to be generated. Thereafter, function structures for each product concept are developed. "A function structure is a set of sub-functions interconnected by flows. Identifying these flows proves effective for helping to partition products into modules. For example, sub-functions with large sets of inter-connecting flows are not good candidates for separation into individual modules" (Dahmus et al. 2001, p. 414).

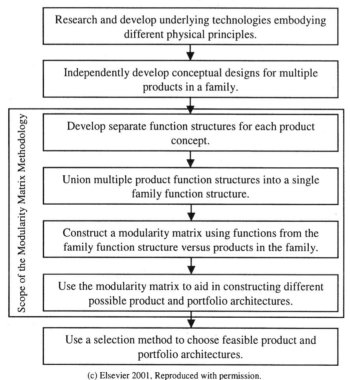

(c) Elsevier 2001, Reproduced with permission.

Figure 8-6. Portfolio architecting process

(Source: Dahmus et al. 2001, p. 413)

After specifying a function structure for each product concept, Dahmus et al. (2001, pp. 416) propose to combine all of these function structures in order to construct a large family function structure that indicates the interrelationships of functions for all of the products in the family. This yields to a single diagram that has every function of every product on it. It is obvious that some products in the family will have common functions. Therefore, these functions are represented only once in the family function structure. Thus, it is possible to have a comprehensive overview of all of the required functions in the family and to make the distinction between the functions that are common to many products and those that are unique to one single product. Dahmus et al. (2001, p. 417) note that some functions may be common for many products but with drastically different flows. For example, the "transmit power" function can be common for several electromechanical products in a family. However, it should be distinguished between the products in which this function conveys rotational or linear motion. Therefore, the family function structure is only a tool for the identification of possible and candidate modules in the family of products.

In order to identify the modules' partitioning, Dahmus et al. (2001, pp. 417) use a set of three heuristics that can be applied to the function structures of each individual product. These heuristics that were developed by Stone et al. (2000, pp. 15) are the dominant flow, branching flows and conversion-transmission flows. Furthermore, Dahmus et al. (2001, p. 417) use a second set of heuristics that is applied to modularize portfolio function structures. These heuristics are divided into two types, namely: shared functions and unique functions. The possible modules identified through the application of the heuristics are visualized within the modularity matrix. "A modularity matrix lists the possible functions from a family function structure as rows in the matrix, then lists the possible products from that family as columns. Each matrix element contains a value that represents the function specification level required" (Dahmus et al. 2001, p. 418). For each individual product, a module candidate is a grouping of one or several functions that is displayed in the product's column. By examining similarities across columns for one function, possible sharing can be identified. The design teams also have to consider additional alternatives to increase commonality in the product family. When it is impossible to define a shared module for all products of the family, it may be advantageous to define a common module only for a product set in the family.

The product architecting process generally results in the generation of several candidate modularizations that should be evaluated in order to select the most optimal modularity matrix.

5.   SUMMARY

Product modularity is considered to be a necessary requirement for the implementation of mass customization. It enables customers to configure their products according to their needs. Moreover, modularity enables the mass customizer to provide many product variants by limiting the costs of variety. In this chapter, we have described the relevant types of modular architectures that are discussed in the technical literature. Then, we have dealt with the benefits and the limits of modularity. We have mentioned that modularity is an important and powerful concept for product customization. However, it must be handled with caution.

We have also emphasized the close interdependence between the product architecture and organization. The development of products around a modular design has many managerial implications that should be taken into account. Product modularity has an influence on the intra-firm level. The development process has to be organized in such a way that there is no need for intensive communication between the different module design teams. In addition, managers have to elaborate incentives to encourage the reutilization of modules and components in order to achieve the economies of substitution. The input of management should be intensified, especially during the definition of the visible design rules in order to create a framework that ensures the integrity of the products. Furthermore, modularity improves strategic learning and helps identify bottleneck capabilities. On the other hand, modularity has an influence on the inter-firm level, in that it has many implications on the supply chain and especially on the relationship between the firm and its suppliers.

Finally, we have presented two relevant tools that can be implemented for the development of modular architectures. The first concept is the Modular Function Deployment (Ericsson/Erixon 1999) and the second is that of Dahmus et al. (2001). Both concepts can be applied in mass customization for the development of modular designs.

Modularity enables the mass customizer to reduce the costs of variety. However, these complexity costs can be reduced only in the case when the product assortment is based on a few modules. It is worth noting that the variety offered in mass customization can be so large that it necessitates a high number of product modules. That is why managers should be provided with adequate tools for the assessment of the impacts of variety on performance. In the next chapter, we will strive for developing a key metrics system that captures the effects of the variety induced complexity in mass customization.

Chapter 9

KEY METRICS SYSTEM BASED MANAGEMENT TOOL FOR VARIETY STEERING AND COMPLEXITY EVALUATION

In part two of this book, we have presented a comprehensive information system that copes with the internal and external complexity problems in mass customization. The information system consists of two main components, namely an advisory system at the front end and a multi-agent system at the back end. The advisory system captures the objective customers' needs in a fast-paced manner. It adapts dialogs and simplifies the elicitation process so that the customers are not overwhelmed by the large product assortment. The goal of the multi-agent based system is to find product configurations with the best chances to fit particular customer requirements. The module agents that are associated with the module variants are able to provide information about themselves in order to support variety steering decisions.

But at the management level, one would require tools that evaluate the orientation on the objective customers' needs. In this context, the first question is: "does the information system actually reduce the external complexity level by efficiently leading customers to select product variants that correspond to their objective needs?" Furthermore, the information system only makes proposals within the scope of variety steering. The final decisions about eliminating or retaining module variants have to be exclusively made by human managers. Subsequently, the second question is: "should the module variants that are proposed by the information system actually be eliminated?" Moreover, an efficient updating of the production program in mass customization not only deals with the elimination of superfluous product variants, but also with the introduction of new variants that correspond to objective customers' needs. Therefore, the third question is: "does the introduction of the new variants induce high complexity and if yes, to which extent?"

All three questions relate to variety and complexity. Whereas the first question deals with the evaluation of the external complexity, the second and third questions relate to decisions that aim at keeping internal complexity under control. To be able to efficiently answer all of these questions, human managers have to be supported by adequate management tools. In the technical literature, there are some methods for supporting management decisions with respect to complexity. The next section will deal with the evaluation of the suitability of the identified methods.

1.   INSUFFICIENCIES OF CURRENT DECISION-SUPPORTING METHODS IN COPING WITH COMPLEXITY

As discussed in chapter three of this book, complexity in mass customization is mainly induced by product variety. Thus, coping with complexity relates to making optimal decisions with respect to internal as well as to external product variety. In this context, it is important to distinguish between two approaches, which are variety management and variety steering. Variety management includes concepts for increasing component and process commonality levels during a company's operations. The main goal is to master operations complexity and to profit from the advantages of both economies of scale and scope when producing variety. Examples of variety management concepts include strategies such as standardization, part families, building blocks, modular product architectures, and platforms (Wildemann 2003, p. 58-59). In contrast, variety steering assumes that some variety management concepts have already been implemented. Variety steering methods can be defined as the set of concepts that aim at the determination of the product variants to be eliminated or introduced to the production program. The main concern is to provide an optimal achievement potential based on (objective) customers' needs by optimizing internal and external complexities. For an automaker, manufacturing cars around platforms and a modular architecture is a variety management decision. For example, a decision may be to consider an engine as a module. Thus, the main objective is to determine the main building blocks and the interfaces that are necessary for assembly. However, variety steering essentially deals with the retention, elimination or introduction of module variants. In the case of the engine module, this means to make decisions about what engine variants (e.g. Diesel 1.6 l, Diesel 2.0 l) in a product family should be eliminated or introduced by taking into account both complexity perspectives. The variety steering methods or tools frequently addressed in the literature are Pareto analysis, contribution margin accounting and activity based costing.

- *Pareto analysis*

Pareto analysis aims at discovering the product variants that are unimportant and transparent to customers. With the help of a Pareto analysis, Nissan automobiles found out that from the 87 existing steering wheels available, around 17 types accounted for 95% of the total installed. In this case, the 70 steering wheels that only account for 5% are considered to be candidates for eventual elimination (Anderson 1997, p. 45).

- *Contribution margin accounting method*

The first step of this method is to examine the contribution margins of both customers and end product variants with the help of an ABC-Analysis. The second step aims at representing both results of the first step in a product/customer-based portfolio. The critical product/customer combinations, namely BC and CC combinations have to be carefully examined for elimination (Wildemann 2000, pp. 65).

- *Activity Based Costing*

The goal of activity based costing is to fairly allocate the complexity costs arising in terms of overheads to the different product variants. So it is possible to provide a more or less accurate cost calculation for the different product variants. Based on the results of this method, the variants presenting high costs that are not honored by the customers can be selected for eventual elimination (e.g. Braun 1999, pp. 83).

Pareto analysis is a past-oriented method. It assumes that variants that have not been perceived by customers, have to be eliminated if there are no further constraints such as e.g. delivery commitment. However, the customers' needs model introduced in chapter four suggests that the customers' unawareness of their objective needs may lead them to purchase product variants other than what they actually want. Thus, the rationalization of the production program on the basis of Pareto analysis may lead to the elimination of product variants with a high potential to satisfy the objective customers' requirements.

Moreover, although the contribution margin accounting method considers two important perspectives, namely customers and end product variants, it is also based on a classification of contribution margins according to an ABC-analysis. In addition, the computation of contribution margins of end product variants seems to be suitable for serial or mass production but not for mass customization assuming a batch size of one. Even if contribution margins can be accurately computed, the analysis may lead to the elimination of single variants consisting of some components or modules being relevant for the manufacturing of other retained product variants. Therefore, this method is not advantageous for mass customization.

The third method, which is activity based costing has a great potential to compute variety costs. However, it is generally associated with high implementation costs, if it is that accurate calculations have to be made. The challenge is to balance the costs needed for the method itself and the savings resulting from its implementation.

Due to the drawbacks of the existing methods for variety steering, we opt for another solution approach based on key metrics. Before determining which key metrics are suitable, we will prove in the next section the suitability of key metrics as an efficient management tool, on which basis mangers can evaluate the achievement of important strategic goals.

2.   KEY METRICS SYSTEMS AS A SUITABLE MANAGEMENT TOOL

Key metrics are quantitative measurements that provide useful information about measurable facts through aggregation and relativization (Reichmann 2001, pp. 19). They can either be used as information or steering instruments. For example, key metrics provided by annual accounts have informational purposes. They describe the company's development in the past and enable the appreciation of business trends. Steering key metrics are used in connection with predefined goals and indicate to what extent these goals are achieved (Kuepper 2001, pp. 344).

But single key metrics generally have a limited informational value. Therefore, both in theory and practice, there are different approaches for structuring the disordered key metrics in order to build systems that can be utilized for predefined objectives. There are basically two different procedures to build key metrics systems. One can start from a key metric that represents an important business goal and build other key metrics in a logical deductive way in terms of a means-end hierarchy. This procedure leads to the construction of deductive systems as e.g. the DuPont scheme. The second procedure is based on mathematical-statistical methods. The corresponding key metrics system referred to as empirical-inductive is selected according to its ability to satisfy a specific analysis goal such as e.g. insolvency prediction (Perridon/Steiner 1999, p. 555).

In business administration, traditional approaches for the development of key metrics systems basically concentrate on monetary perspectives. In order to avoid this main disadvantage, Kaplan/Norton (1997, pp. 24) have developed the balanced scorecard which structures key metrics according to four perspectives, namely: the innovation and learning, internal business processes, customer, and financial perspectives. The balanced scorecard should include 15 to maximally 25 key metrics that are important to efficiently control the implementation of the firm's strategy. Since the balanced scorecard focuses on strategically relevant key metrics and points out the cause-effects relationships between the different perspectives, it can be considered as a management instrument which supports leaders in evaluating the goals' achievement with respect to the strategy.

Weber/Schaeffer (1999, pp. 11) argue the relevance of the balanced scorecard as a well-structured management tool. However, they note that the balanced scorecard should be considered as a diagnostic and not as an interactive control system. In this context, the number of key metrics (15-25) suggested by Kaplan/Norton (1997) in the balanced scorecard is too large to enable an efficient interactive and permanent control by managers. According to Miller (1956), the capacity of humans to simultaneously receive, process and remember simultaneously incoming data is limited to seven units (plus or minus two). On this basis, Weber/Schaeffer (1999, p. 13) conclude that managers can efficiently concentrate on approximately seven key metrics for an ongoing control. This is in accordance with the concept of the selective key metrics system of Weber et al. (1995, pp. 9) who argue that a focused system consisting of only a few key metrics is required to permanently and closely monitor critical bottlenecks in business. It is worth noting that both concepts of a balanced scorecard and selective key metrics are not incompatible. They are complementary because the selective system should be a subsystem of the balanced scorecard. It only consists of the key metrics that are very critical for competitiveness.

Key metrics are also more often discussed in connection with performance measurement (e.g. Maskell 1991). Performance measurement is "...the process of quantifying the efficiency and effectiveness of action", whereas a performance measure is "...a metric used to quantify the efficiency and/or effectiveness of an action" (Neely et al. 1995, p. 80). "A performance measurement system can be defined as the set of metrics used to quantify both the efficiency and effectiveness of actions" (Neely et al. 1995, p. 81). Performance measures encourage the strategy's implementation, which is considered to be a very important management task. They also enhance continuous improvement because they enable the evaluation of important performance dimensions. Furthermore, it is widely accepted that performance measures influence behavior inside the company.

"Measures that are aligned with strategy not only provide information on whether the strategy is being implemented, but also encourage behaviours consistent with the strategy" (Neely 1999, p. 212).

A performance measurement system can be seen as a part of a wider system called the strategic control system, which additionally includes goal setting, feedback, and reward or sanction (Neely et al. 1995, p. 102). Very important issues which are more often discussed in academia and practice mainly concern (a) what should be measured, (b) which metrics have to be drawn on in a specific case, and (c) what techniques can be used to reduce the list of possible measures to a meaningful set that can be easily monitored by managers.

Key metrics systems are suitable management tools because they provide a short and concise description of difficult, complex matters of facts (Mueller 2001, p. 46). Following the discussion that was elaborated above, key metrics should also be correctly and error-free designed with respect to predefined objectives. They are relevant to monitor and control the implementation of the company's strategy. As aforementioned, the mass customization strategy aims at achieving a high individualization level nearly at mass production efficiency. However, the most important problem that may hinder a successful implementation is complexity due to a large product variety. Therefore, managers require a management tool in order to be able to efficiently quantify and detect the internal and external complexity problems induced by product variety. Due to the suitability of key metrics systems, we will strive for the development of a system capable of detecting complexity problems.

3.   A SUB-PROCESS MODEL FOR THE DETERMINATION OF THE KEY METRICS

In order to derive the important key metrics for complexity evaluation, an analysis of the most relevant sub-processes in mass customization should be carried out. For this reason, we draw on the main internal abilities of a mass customizing system that are determined in section 4.2 of chapter two. Figure 9-1 shows the correspondence between the internal abilities and the sub-processes. The main sub-processes in mass customization are: the interaction, development, production, purchasing, logistics and information sub-processes.

Figure 9-1. Correspondence between internal abilities and sub-processes

Note that variety and complexity management have no correspondent sub-processes. However, there are mutual interactions between variety and the different sub-processes excepting the information sub-process which is considered as a cross-sectional process that ensures a smooth information flow between all other sub-processes. With respect to variety steering, the sub-process analysis has many advantages because:

- it is comprehensive and encloses all the variety driving activities;
- it provides an efficient methodology to structure the variety problem;
- it shows that decisions related to variety at one sub-process level influences the performance of other sub-processes.

Product variety can considerably influence the interaction process. As the number of product variants increases, the risk that customers can be confused and overwhelmed by the mass customizer's offer also increases. In this case, the interaction process in which customers configure products on the basis of modules is no longer efficient and the logic of the interaction process should be based on an advisory as discussed in chapter five of this book. In other words, only the product variants with the best chances to respond to the objective needs should be displayed to the customers.

Conversely, the interaction process may have an influence on product variety. If the product variants matching the objective needs are not recognized during the interaction process then variety rationalization efforts aiming at reducing complexity through the elimination of superfluous product configurations would suggest the retention of variants that correspond to the subjective needs. This will lead to suboptimal product assortments that are not oriented on the objective needs. But if the product variants matching the objective needs are recognized, the product assortment can be better optimized. Thus, the mutual interactions between variety and the interaction sub-process are obvious.

The development, purchasing, production and logistic sub-processes have a direct influence on variety inside a company. It is important to emphasize that variety affects other aspects such as e.g. suppliers, interfaces, quality etc. In the development sub-process, new models, product variants, modules or components are introduced by engineers, which can increase the internal and external product variety, or both. Furthermore, the vertical range of manufacture is becoming lower because the components that do not correspond to core competencies are generally outsourced. As a result, the purchasing sub-process has a great potential to increase component variety. Suppliers' variety can also be influenced and the outsourcing strategy of the company widely determines the extent of component part numbers. In practice, it is not rare to observe that a standard component with no strategic significance has many different part numbers and is purchased from different suppliers, which represents a considerable loss of efficiency.

In the production sub-process new product variants may require e.g. new machine tools that can either be manufactured within the company or outsourced. The logistics sub-process can also increase product variety. Adaptations according to particular country prescriptions can be carried out at a late point in time in the supply chain, for example during the distribution process.

The possible effects of product variety on distribution, production, purchasing and research and development are qualitatively scrutinized by Lingnau (1994, pp. 127-148). Lingnau also examines the cost effects due to increasing product variety in each sub-process. For example, the introduction of new product variants may necessitate new distribution channels, which are associated with additional costs. Moreover, longer idle times due to frequent setups can be triggered. Increasing variety also means steadily changing bottlenecks in production, so that additional costs are incurred because of idle manufacturing equipment. Because of higher variety the work-in-process inventory can considerably increase and quality assurance measures must be intensified. Figure 9-2 summarizes the interactions between the different sub-processes and product variety. In the following, we more closely examine each sub-process.

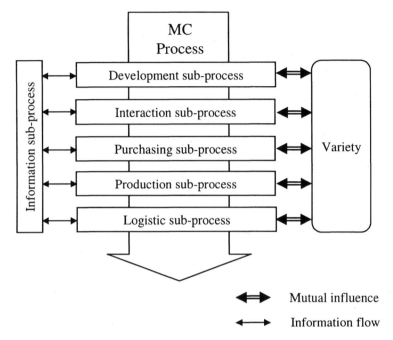

Figure 9-2. Relevant sub-processes in mass customization

3.1   Development Sub-process

The product development stage determines 80 percent of the lifetime cumulative product cost (Anderson 1997, p. 131; Anderson 2004, p. 57). Especially in mass customization, product development gains more in importance because costs should be kept low in spite of the high individualization level. The product architecture to a great extent impacts the scope and costs of product variety (Ulrich/Eppinger 2000, p. 186). It determines 60 percent of a product's cost and presents a high leverage opportunity for reducing costs (Anderson 1997, pp. 132; Anderson 2004, p. 57).

Product architecture can either be modular or integral. A modular architecture involves a low number of functions per component, whereas integral architecture presents a higher integration of functions per component. Integral product architectures trigger high complexity costs during the product life cycle (Ishii 1998, p. 3). Ulrich/Eppinger (2000, p. 186) point out that modular chunks allow changes to be made to a few isolated functional elements of the product without affecting the design of other chunks. However, when altering an integral chunk, changes are required to be made to other related chunks.

The ideal product architecture for mass customization should enable the manufacturing of a high number of product variants on the basis of a high commonality of parts and components. The product platform concept plays a decisive role and presents an efficient means to reduce complexity. A product platform is generally understood as a set of parts or components, which establish a common structure for many products. However, platform strategies may not only include the shared components across many product variants but also the manufacturing processes and knowledge required for the production of variety (Piller/Waringer 1999, pp. 64; Ulrich/Eppinger 2000, p. 200; Siddique/Rosen 2001, p. 1).

3.2   Interaction Sub-process

The main goal of the interaction process is to obtain relevant information from the customer that enables the configuration and manufacturing of individualized products (Piller 2000, pp. 195). Thus, customers are integrated into the value adding process and are considered as "co-producers" or "prosumers" (Toffler 1980, p. 275; Piller 2000, p. 196). The extent of customer integration may vary from the simple configuration on the basis of predefined modules and components (e.g. when the mass customizer is just an assembler) to the co-design of products (e.g. when the mass customizer is a fabricator).

The interaction activity can either be supported by trade or sales or carried out directly over the Internet. The interaction process is a critical part of the mass customization service process and influences the overall perception of quality. The customer satisfaction level depends not only on the end product quality, but also on the interaction process. A lack of transparency during interaction may confuse customers and lead to process abortion (Riemer/Totz 2001, p. 5).

3.3   Purchasing Sub-Process

The decisions made at the development stage with respect to components' variety have a great impact on purchasing. During this sub-process, the main complexity drivers are parts and material variety, suppliers' variety, interfaces' variety and quality variety. In order to optimize the purchasing process, Wildemann (2000, p. 33) suggests carrying out an ABC-analysis with regard to purchasing volumes of the different material groups. For A-materials, the purchasing processes should be carefully examined. However, for B and C material groups, it is advantageous to opt for a few typical processes with a relatively low complexity level.

In addition, dealing with a huge number of suppliers triggers high complexity, hinders cooperative relationships and negatively affects quality. By reducing the number of suppliers, an increasing efficiency of purchasing can be attained. For instance, the concept of single sourcing suggests for one to retain just one supplier for a specific part or parts' family over a long period of time. This enables the achievement of advantages concerning the costs, quality and delivery time levels (Maskell 1991, p. 210; Wildemann 1997, pp. 87).

The external purchasing of modules respectively systems is said to be modular respectively system sourcing. Both concepts combine conflicting goals of lowering the vertical range of manufacturing and reducing the number of suppliers (Wildemann 1997, p. 90; Wildemann 2000, p. 39). In mass customization, modular and/or system sourcing lead to a considerable reduction of purchasing complexity at the process and component levels.

3.4   Production Sub-Process

The production process plays an important role in the success or failure of mass customization. In order to be able to offer a high product variety while maintaining a competitive customer service time, the manufacturing system should be flexible (Kaluza 1989, pp. 287). Flexibility is guaranteed when the delays that arise from switching over from one variant to another are considerably reduced, ideally to zero (Anderson 1997, p. 177).

A relevant concept aiming at reducing complexity in production is to apply modularity on the shop floor. The resulting organization is a modular one based on manufacturing cells "...where dissimilar machines are grouped together based upon the manufacturing process that is completed within the cell" (Maskell 1991, p. 157). Furthermore, modular organization reduces the number of linkages between machines and leads to short production lead times (Wildemann 2000, p. 47).

In addition to commonality at the components level, the mass customizer has to strive for a high production process commonality, which means that the different variants can be manufactured on the basis of a few production processes. As a result, a certain level of stability as well as straightforwardness is achieved (Maskell 1991, p. 157). Thereby, production complexity and perturbation sensitivity of manufacturing processes are considerably reduced (Wildemann 1997, p. 152). Honeywell's thermostat production facility in Golden Valley, Minnesota provides a good example, which illustrates the importance of flexible manufacturing by increasing process commonality. This company originally had different production lines for each of its three products. By means of a flexible manufacturing system, all of their products are manufactured on a single line, while product changeover time is slashed from 25 minutes to three minutes (Berman 2002, p. 56).

3.5   Logistic Sub-Process

The design of the logistic system in mass customization can provide some additional individualization opportunities. For instance, the customer may choose from many logistic options related to payment, packaging and transport. Real individualization relates to two main aspects, namely individual packaging (e.g. gift-wrapping and packages enclosing individual greetings) and individual delivery times. The mass customizer attains a real competitive advantage when the individual time schedules of customers are respected (Riemer/Totz 2001, p. 6).

Mass customization does not generate inventories at the end product level. However, work-in-process inventories of mass-produced modules and standard components, as well as raw materials for individualized processing can exist. To check out whether the logistic system of a mass customizer is smoothly working, the work-in-process inventory must be permanently kept in view. In fact, "[m]ost things that go wrong in a logistics system cause inventory to increase" (Tersine/Wacker 2000, p. 114).

To improve the performance of logistics, the just-in-time concept provides interesting clues. Three main requisites should be available in order to adequately implement the just-in-time concept. They are: (a) a pull system (e.g. Kanban) for production control, (b) small batch sizes and reduced setup times and (c) stable and reliable production operations. According to this, mass customization and just-in-time are not conflicting approaches. Just-in-time seems to be a relevant requisite for mass customization in order to ensure an increasing of flexibility while decreasing work-in-process inventories (Wildemann 1995a, pp. 112).

In addition, mass customization requires a highly coordinated supply chain. Customers must be provided with the right goods at the right time. A virtual integration enables the mass customizing company to maintain close relationships with partners that have excellent capabilities for performing specific functions (Berman 2002, p. 56). The main advantage of such integration is to benefit from partner specialization, extend the individualization level and increase flexibility while reducing the internal complexity (Rogoll/Piller 2002, pp. 24).

3.6   Information Sub-Process

In mass customization, before receiving an order, it is impossible to exactly determine the required parts and products that must be manufactured and shipped. In order to ensure an integrated information flow from order taking to delivery, the information sub-process plays a relevant role. An effective and efficient information system for mass customization should capture the product configuration of the customer, develop a list of product requirements, determine manufacturing specifications with respect to routing, material processing, assembly, etc. The information system should also offer the possibility to set up the manufacturing system, arrange for end product shipment and enable the verification of a product's order status (Berman 2002, pp. 57). It is noteworthy that with the increasing number of product variants the information process complexity exponentially increases.

Information systems for mass customization should enable the supply chain to operate as an integrated unit. The integration of the different information systems guarantees that the whole system runs effectively. For example, when an unexpected change arises, the suppliers can immediately react and adjust their activities. This leads to an increasing agility, which improves the system's abilities to adapt to unforeseen events (Oleson 1998, p. 91). In this context, Turowski (2002, pp. 69) emphasizes that automated inter-company communications are necessary and that e.g. an agent-based system can automate procurement and inter-company coordination of production.

4. COMPLEXITY KEY METRICS

On the basis of the described sub-process model, the relevant key metrics for the detection of variety-induced complexity in mass customization will be derived. To guarantee a certain computational ease of the key metrics, those related to costs will not be drawn on. In effect, calculating accurate complexity costs by applying activity based costing is associated with enormous expenses. Furthermore, activity based costing may lead to inaccurate cost estimation (Schaefer 1993, pp. 311; Eberle 2000, p. 346).

For the determination of the key metrics, it is assumed that no activity based costing system is implemented. The extent of variety-induced complexity is captured on the basis of non-cost based measurements which are related to e.g. time, decoupling point, etc. Moreover, the key metrics have to be based on data already available in the company. They should provide valuable information for the evaluation of complexity without requiring additional efforts for e.g. data collection.

4.1   Complexity Key Metrics for the Development Sub-Process

During the development process, engineers have to consider already devised and used parts. The main goal is to use each part in as many products as possible because *"products that use many common parts inherently have less variety cost than products with unique parts"* (Anderson 1997, p. 78). The development of new products on the basis of standardized and common parts reduces complexity at the development stage. The use of unique parts has to be kept at a low level, ideally at zero (Maskell 1991, p. 178; Anderson 1997, p. 92). Martin and Ishii (1997, p. 3) have defined the commonality index (CI) indicating to what extent the different product variants within a product family are based on fewer unique parts.

$$[1] \qquad CI = 1 - \frac{u - \max p_j}{\displaystyle\sum_{j=1}^{v_n} p_j - \max p_j}$$

$$0 < CI \leq 1$$

u	Number of unique part numbers
p_j	Number of part numbers in model j
v_n	Final number of varieties offered

(Source: Martin/Ishii 1997, p. 3)

As aforementioned, "[t]he best method for achieving mass customization … is by creating modular components that can be configured into a wide variety of end products" (Pine 1993, p. 196). However, formula [1] seems to be unsuitable for products developed around modular architectures. If all product variants result from the combination of a finite number of modules, then it is legitimate to set u=0, because one would consider the modules as common parts. Subsequently CI<0, which is totally absurd. In their research, Martin/Ishii (1997) only indicate the formula, but do not explain how to determine the unique parts. Therefore, the main concern should be to determine the criteria according to which module variant is considered as common or unique.

Maskell (1991, p. 179) notes that the method to be used for the distinction between common and unique parts can be different from one company to another. In general, the product parts built in a small number of variants are non-common parts. However, the components built in a large number of end product variants are not usually common. If the product variants represent a small percentage of the company's output, then the corresponding parts should be considered non-common parts.

For the analysis of commonality, Anderson (1997, pp. 103) considers the part usage volume as a measurement for the determination of the "high runners" and "low runners" in a category of parts. Whereas high runners have a high usage volume, low runners have a low usage volume. In addition, Anderson (1997) suggests determining the number of different product variants, in which the part is built in, for each part. Both results must be plotted in one graphic. The x-axis relates to the product part numbers, which are ordered in such a way that the usage level is strictly decreasing (Pareto order). The usage volume is represented above the x-axis, whereas the number of products is plotted under the x-axis. Thus, usage patterns for both criteria can be simultaneously analyzed. The common parts are those that have a high usage level and are built in a wide number of product variants. This method can be applied when dealing with the rationalization of modules and parts lists.

The combination of both common part and module strategies leads to the common modules strategy (Nilles 2002, p. 140). A high commonality at the module level should not involve low parts commonality within- and between- modules. The modules' strategy reduces product complexity but the modules complexity itself has to be kept low. That is why key metrics that capture commonality at the modules and parts levels are of value. Obviously, the parts that are built in outsourced modules are not considered when computing commonality. Only those parts that are assembled during the company's operation should be taken into account. Blecker et al. (2003a, pp. 16-17) propose two key metrics (modules commonality metric [2] and parts commonality metric [3]) in order to separately track both of the commonality levels of mass customized products. The modules (parts) commonality metric is defined as the quotient between the number of common modules (parts) and the number of all modules (parts).

[2]
$$MCM = \frac{\text{Number of common modules}}{\text{Number of all modules}}$$
MCM Modules commonality metric

[3]
$$PCM = \frac{\text{Number of common parts}}{\text{Number of all parts}}$$
PCM Parts commonality metric

For the computation of the number of common modules, Blecker et al. (2003a, pp. 16-17) propose an algorithm which takes into account both criteria of Anderson (1997, p. 104), namely the usage level and number of products using the module or the part. The algorithm is proposed for modules and can also be adapted for parts.

- For each module M_i, determine the product variants P_j which have already used the module M_i
- Determine the (average) number n_{ij} of module M_i assembled in P_j
- Determine the sales volume in unit S_j of P_j
- For each module M_i compute the term L_i, where

[4]
$$L_i = \frac{\sum\limits_{j=1}^{k_i} n_{ij}\, S_j}{\sum\limits_{i=1}^{nm}\sum\limits_{j=1}^{k_i} n_{ij}\, S_j}$$

nm Number of all modules

k_i Number of products using the module M_i

- For each M_i
  If ($L_i - 1/nm$)>0 then M_i is a common module
  If ($L_i - 1/nm$)<0 then M_i is not a common module

(Source: Blecker et al. 2003a, pp. 16-17)

Collier (1981, pp. 86; 1982, pp. 1296) suggests an analytical measurement for product structure called the degree of commonality index (C) which can be defined for a single end item, a product family, an entire product line or for any level of the product structure. The index represented by formula [5] captures the average number of common parent items per average distinct component part, which means to count the mean number of applications per component.

Degree of Commonality Index, $C = \dfrac{\sum\limits_{j=i+1}^{i+d} \Phi_j}{d}$ $1 \le C \le \beta$

Φ_j Number of immediate parents for component j

[5] d Total number of distinct components in the set of end items

     i Total number of end items

     β Total number of immediate parents for all components

         in the set of end items ($\beta = \sum\limits_{j=i+1}^{i+d} \Phi_j$)

(Source: Collier 1981, pp. 86-87)

In order to illustrate how to use the degree of commonality index by Collier (1981; 1982), we consider the following example represented by figure 9-3. The product structure consists of $i = 2$ end items and $d = 5$ distinct components in the set of end items. In order to compute the degree of commonality index, one has to firstly count the number of immediate parents for each component. So, component 3 has two immediate parents, which are 1 and 2, whereas component 4 only has one immediate parent which is component 1. Component 5, 6 and 7 have respectively three, two and three immediate parents. The degree of commonality index (C) is obtained by dividing (2+1+3+2+3=11) through the number of different components excepting the end items (d=5). Subsequently, C=2.2.

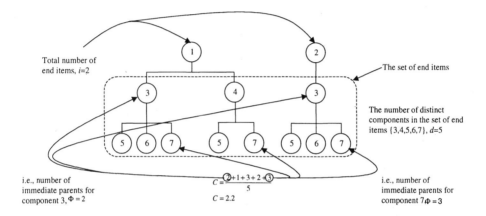

Figure 9-3. Example for the calculation of the degree of commonality index
(Adapted from: Collier 1982, p. 1297; Siddique 2000, p. 82)

For the evaluation of commonality, Jiao/Tseng (2000, pp. 229) elaborate a key metric $CI^{(C)}$ called the component part commonality index. In addition to the mean number of applications per component proposed by Collier (1981; 1982), the index takes into account the product volume, quantity per operation and price/cost of the component part.

In order to compute the key metric, a bill of material-like product structure should be employed for the representation of the product family as illustrated by figure 9-3. This product structure has a tree form with many hierarchical levels (parent-child relationships). Each level contains one or more nodes that indicate product components or end items. The highest-level node represents an end item, whereas the lowest-level nodes, the purchased parts. The intermediate nodes represent manufactured or assembled parts. Whereas a one-to-one parent-child relationship relates to a manufacturing operation, a one-to-many relationship is an assembly operation.

Additionally, it is important that the product structure tree specifies the number or quantity of components required for the realization of each operation (Jiao/Tseng 2000, p. 229).

It is noteworthy that formula [6] is a corrected version of the formula proposed by Jiao/Tseng (2000, p. 232) in order to ensure mathematical rightness, which is necessary for an error-free computation. Indeed, the index ij has been assigned to the term $\sum_{h=1}^{n_h}\prod_{k=0}^{n_k} q_{hk}$, which depends on both the product i within the family and the component j.

[6]
$$CI^{(C)} = \frac{\sum_{j=1}^{d}\left\{P_j\sum_{i=1}^{m}\Phi_{ij}\sum_{i=1}^{m}\left[V_i\left(\sum_{h=1}^{n_h}\prod_{k=0}^{n_k} q_{hk}\right)_{ij}\right]\right\}}{\sum_{j=1}^{d}\left\{P_j\sum_{i=1}^{m}\left[V_i\left(\sum_{h=1}^{n_h}\prod_{k=0}^{n_k} q_{hk}\right)_{ij}\right]\right\}} \quad 1 \le CI^{(c)} \le \alpha$$

$$\alpha = \sum_{j=1}^{d}\sum_{i=1}^{m}\Phi_{ij} \ge 1$$

(Source: Jiao/Tseng 2000, p. 232)

d The total number of distinct component parts used in all the product structures of a product family

m The total number of end products in a product family

i The index of each member product of a product family $i = 1,\ldots,m$

j The index of each distinct component part d_j, $j = 1,\ldots,d$

P_j The price of each type of purchased parts or the estimated cost of each internally made component part

Φ_{ij} The number of immediate parents for each distinct component part d_j over all the product levels of product i of the family

h	A particular path from the item d_j to the end item node through the levels of the product tree by including node d_j and excluding the end item node
k	The index of the nodes on path h
n_h	Total number of paths for d_j within product i
n_k	Total number of parent nodes on path h
k	The index of the nodes on path h, $k = 1,...n_k$ ($k = 0$ represents the node d_j itself.)
q_{hk}	The quantity per operation (either manufacturing or assembly) of node k required by its immediate parent node along path h
V_i	The volume of end product i in the family

(Source: Jiao/Tseng 2000, pp. 229)

According to the key metric developed by Jiao/Tseng (2000, p. 232), the commonality index increases when a few parts are used across a high number of product variations in the product family. The commonality metric also depends on the sales volume of each product variation. If engineers increase the use of a component assembled in rarely sold product variations, the commonality index will not increase considerably. As a managerial implication, it is more beneficial to increase components utilization inside product variants with high sales volumes. The value of the index also increases with the quantity or number of components required for the manufacture of a product. Another variable which influences the commonality index is the price of the component if it is outsourced or its cost if it is manufactured during a company's operations. The utilization increase of high price/cost components across the product family considerably increases the part commonality in comparison to the low price/cost components. However, "...the costs of internally made parts are not straightforward to obtain" (Jiao/Tseng 2000, p. 230). Therefore, Jiao/Tseng (1999, pp. 742) propose a pragmatic approach to cost estimation based on standard time estimation. This approach enables managers to already rapidly appreciate product costs at the design phase. The idea is that overall overheads can be fairly assigned to the products by assuming a linear correlation between standard times needed for the execution of manufacturing activities and overheads. Thus, the analysis of standard times required during operations enables the derivation of overhead costs for each new product. The authors have reported that according to their experience with respect to a case study, the approach provides results with a maximum cost deviation from the actual cost of twelve percent. This pragmatic method shows promise of reducing some drawbacks of traditional approaches to product costing. However, there is no empirical evidence demonstrating that the method will generally provide good results among many industries.

Recapitulating, several key metrics and methods for the determination of component commonality are presented above. The main question is "among all identified key metrics, what is the most suitable key metric for the evaluation of commonality?" We argue that commonality is a very important metric that has to be tracked by the mass customizer in order to be able to capture the proliferation of internal product variety. But the suitability of the key metric largely depends on the company's requirements. The key metric of Collier (1981; 1982) is easy to compute, but does not consider relevant aspects such as product volume or quantity of components per operation, etc. The key metric proposed by Blecker et al. (2003a) takes into account the number of parts required per operation as well as the sales volume, but does not consider e.g. the average application of the component in a product. The most comprehensive key metric is proposed by Jiao/Tseng (2000). The difficulty that can be encountered when computing this key metric may be the determination of part costs. Nevertheless, the key metric can be adapted by eliminating the term related to costs/prices in formula [6], so that the index only depends on the average application of components, product sales volume and quantity per operation. We obtain the following index [7].

[7]
$$CI^{(C)} = \frac{\sum_{j=1}^{d}\sum_{i=1}^{m}\Phi_{ij}\sum_{i=1}^{m}\left[V_i\left(\sum_{h=1}^{n_h}\prod_{k=0}^{n_k}q_{hk}\right)\right]_{ij}}{\sum_{j=1}^{d}\sum_{i=1}^{m}\left[V_i\left(\sum_{h=1}^{n_h}\prod_{k=0}^{n_k}q_{hk}\right)\right]_{ij}} \quad 1 \le CI^{(c)} \le \alpha$$

$$\alpha = \sum_{j=1}^{d}\sum_{i=1}^{m}\Phi_{ij} \ge 1$$

In order to improve commonality in a product family, it is advantageous to create a common basis for many derivative products and models by developing a product platform. Nilles (2002, p. 136) defines a product platform as a spatially closed functional unit with well-defined interfaces. In addition, it represents a standardized part of the product system structure. However, one has to make the distinction between product platforms and standardization. "Standardization and product platforms are not synonymous; standardization can be achieved through robust platforms, but robust platforms themselves cannot be achieved solely through standardization" (Meyer/Lehnerd 1997, p. 120).

Meyer/Lehnerd (1997, p. 147) mention that traditional measures are seldom useful for the assessment of platforms and product families' performance because these measures mainly focus on single products and projects. Meyer/Lehnerd (1997, p. 160) introduce the platform cycle time efficiency and point out that overall efficiency of the product platform increases when follow-on products can be rapidly created without enormous efforts and costs. "A strong product platform, while taking relatively longer to create than derivative products, should allow the firm to experience rapid generation of those derivatives" (Meyer/Lenherd 1997, p. 159). Subsequently, the complexity of the development process decreases considerably because different product variants can be easily and efficiently derived. Furthermore, the lower the value of the key metric is, the better.

$$PCTE = \frac{\text{Elapsed time to develop a derivative product}}{\text{Elapsed time to develop the product platform}}$$

[8]

$PCTE$ Platform Cycle Time Efficiency

(Source: Meyer/Lehnerd 1997, p. 160)

PCTE is defined for a particular derivative product. But it is possible to define an average cycle time efficiency value for a generation of products based on a particular version of a product platform as described by formula [9] (Meyer/Lehnerd 1997, p. 160).

$$APCTE = \frac{\text{Average elapsed time to develop a derivative product}}{\text{Elapsed time to develop the product platform}}$$

[9]

$APCTE$ Average Platform Cycle Time Efficiency

(Source: Meyer/Lehnerd 1997, p. 160)

According to a case study showing an application of the key metric [9], Meyer/Lehnerd (1997, p. 160-161) point out that design engineers have to be provided with enough time and resources for the development of the product platform. Otherwise, the development of derivative products will require longer times, which involves poor efficiency. The platform development efforts should not be treated as if they were derivative product developments. When a platform is released too early because of e.g. competitive pressure, the same platform will rather act as a barrier to rapid development of further product variations, than flexibility and time-to-market enhancement factors.

Meyer/Lehnerd (1997, pp. 152) have developed other key metrics in order to evaluate product platforms performance (e.g. platform effectiveness, cost price ratio). However, we argue that among all of the key metrics proposed, both key metrics [8] and [9] are the most suitable for the evaluation of complexity in the design process. For example, if the value of the cycle time efficiency is very low (e.g. in comparison to past values or industrial branch values gained from external Benchmarking), this would suggest that the introduction of new product variants would be relatively simple and not cost-intensive. However, when the value of the cycle time efficiency is high, the derivation of new products would involve complexity at the design process because many adaptations at the platform level and the derivatives may be necessary.

Platform efficiency provides useful information about the ease in deriving new products, but it does not provide managers with insights into product architecture flexibility. In mass customization, it is advantageous to design the product architecture, so that a large number of product variants can be derived on the basis of a few modules. For this reason, we draw on the multiple use metric (E_v) of Ericsson/Erixon (1999, p. 127). This key metric is computed as the quotient between the number of product variants and the total number of required modules. A high value of the multiple use metric indicates that the whole range of product variants can be produced on the basis of a few modules. For example, the panel meter of Nippondenso can be assembled into 288 variants out of 16 total modules and has an E_v of 18 (Ericsson/Erixon 1999, p. 127).

[10] $$E_v = \frac{N_v}{N_{mt}}$$

E_v Multiple use metric

N_v Number of product variants required by customers

N_{mt} Total number of modules required to build up all the

            product variants

(Source: Ericsson/Erixon 1999, p. 127)

A necessary condition for the generation of a wide range of variety by mixing-and-matching a few modules is an optimal design of module interfaces. In fact, the main complexity of modular architectures relates to the specification and standardization of the interfaces (Mikkola/Gassmann 2003, pp. 208). "The specification of an interface is defined by gathering data on items such as form, fixation principles, number of contact surfaces and attachments, number of energy connection points, material flow, and signals. A certain amount of information is needed for every interface. Low complexity is, in this case, synonymous with ease to specify (low information content)" (Ericsson/Erixon 1999, p. 108). A low interface complexity enables more efficient concurrent engineering where modules can be simultaneously developed by different R&D teams. As a result, development lead times drastically decrease and products can be launched onto the market quicker (e.g. Ericsson/Erixon 1999, pp. 17, Wildemann 2003, p. 157). The interface complexity metric is defined as follows:

$$[11] \quad I_c = \frac{\sum_{i=1}^{N_m-1} T_i}{A_t}$$

I_c Interface complexity metric

N_m Number of modules in one product variant

T_i Assembly time for one interface

A_t Ideal assembly operation time

(Source: Ericsson/Erixon 1999, p. 109)

4.2 Complexity Key Metrics for the Interaction Sub-Process

The interaction between the supplier and customers is a relevant key sub-process in mass customization. It determines to a great extent the success or failure of the whole customization process. The interaction process fails if it does not efficiently help customers recognize their optimal product variants. Consequently, customers may order and receive product variants that do not correspond to their real needs, which triggers disappointment and mistrust with respect to the benefits of mass customization.

Customers may experience high complexity during configuration because of large product variety. Although the ability to manufacture a high variety range is relevant for the fulfillment of different customers' needs, it is not usually obvious, whether customers actually require and honor this variety. Thus, the mass customizer only has to offer an achievement potential that maximizes customer satisfaction while retaining low costs. In order to track the optimality level of the assortment from the customer's perspective, we draw on the key metric referred to as "used variety" which captures the perceived variety in comparison to the theoretically possible product variants. Low values of this metric indicate that a high number of product variants are unperceived or uninteresting from the customers' perspective.

$$UVM = \frac{\text{Number of perceived variants}}{\text{Number of all possible variants}}$$

[12]

UVM Used Variety Metric

(Source: Piller 2002, p. 15)

When speaking about complexity from the customer's perspective, it is relevant to make the distinction between perceived and actual complexity. Customers experience difficulties during the decision making process because of the perceived complexity, but not necessarily because of the actual complexity or variety. Actual variety may be small, but customers may perceive a high variety (e.g. the Chinese menu which seems to offer a high number of choices but actually based on four kinds of meat, sauces, etc.). The implementation of an advisory system decreases the perceived complexity, while the actual complexity can be very high. Thus, both types of complexity are independent from each other. As a managerial implication, the mass customizer should not necessarily increase product variety in order to improve the level of variety that is perceived by customers. In this context, the information presentation format plays a decisive role because it facilitates information absorption and decreases uncertainty during the decision making process (Huffman/Kahn 1998, p. 493). However, if customers are overloaded with information, they would not complete and abort the interaction process (Piller/Tseng 2003, p. 519).

Besides the used variety, two other key metrics are proposed in order to capture the perceived complexity level, namely the average time required to select a product variant and the interaction abortion rate. Piller (2000, p. 279) notes that the configuration of product variants should be completed in a few minutes, in a few hours when the product is considerably complex, and never in several weeks. High abortion rates and long interaction times would suggest the unsuitability of the presentation format for displaying product variety. When analyzing the average interaction time, it may be useful to distinguish between old and new customers because old customers may take less time to select a product variant than new customers.

$$CT = \frac{\sum_{i=1}^{N} CT_i}{N}$$

[13]

CT Average interaction length of time

CT_i Time needed from one customer to fulfil one configuration

N Number of fulfilled configurations

$$AR = \frac{\text{Number of aborted interaction processes}}{\text{Number of log - ins}}$$

[14]

AR Abortion rate

4.3 Complexity Key Metrics for the Purchasing Sub-Process

The mass customizer can reduce complexity of the purchasing process by applying concepts such as process standardization and modular sourcing. Furthermore, commonality at the product level reduces the number of unique parts whose procurement is associated with high costs and long delivery times. Anderson (1997, pp. 93; 2004, pp. 151) points out that it is advantageous to examine, whether it is possible to replace the part having the optimal size with the part having the next larger standard size. Although this may involve higher direct costs, cost advantages due to savings at the overhead costs may be larger. In order to track the purchasing process complexity, we draw on the purchasing commonality metric defined by formula [15]. This metric measures the number of common purchasing processes in comparison to the number of all of the purchasing processes.

$$PPC = \frac{\text{Number of common purchasing processes}}{\text{Number of all purchasing processes}}$$

[15]

PPC Purchasing Process Commonality (within a part category)

4.4 Key Metrics for the Production Sub-Process

Offering acceptable delivery times in mass customization is a great challenge that presupposes the high capabilities of the manufacturing system. Before the customer order arrives, it is impossible to predict which individualized variants have to be manufactured. However, mass-produced components and subassemblies can be independently stored from the customer order. The variant determination point should be placed at the end of the value chain in order to avoid early variety proliferation. So it is in fact possible to optimize inventory costs while offering high delivery service (Wildemann 1995b, pp. 190, Waller et al. 2000, p. 134).

Martin/Ishii (1996, p. 6) define the differentiation point index *DI* that captures the position where the product differentiation occurs within the process. The denominator of this index shows the worst case where all of the variants are determined at the beginning of the production process and the numerator reflects to what extent the actual process flow has moved away from the worst-case situation. Thus, low values of *DI* reflect good capabilities of the production process to cope with variety-induced complexity.

[16] $$DI = \frac{\sum\limits_{i=1}^{n} d_i\, v_i\, a_i}{n\, d_1\, v_n \sum\limits_{i=1}^{n} a_i}$$

v_i Number of different products exciting process i

n Number of processes

v_n Final number of varieties offered

d_i Average throughput time from process i to sale

d_1 Average throughput time from beginning of production to sale

a_i Value added at process i

(Source: Martin/Ishii 1996, p. 6)

This index has a high relevance in mass customization. It incorporates all of the parameters (lead times, value growth along the entire process and the number of variants), which according to Wildemann (2000, p. 47) are crucial for the determination of the optimal differentiation point. The introduction of new product variants may affect the value of the index. Therefore, before adding product variants to the production program, the value of the index should be carefully examined. In the case when the index remains constant or varies a little, the overhead costs due to inventory will not significantly increase. If not, it is relevant to examine whether it is profitable to introduce these new product variants. Figure 9-4 shows two production processes with two differentiation indexes. The left side of the figure shows a bad situation where the proliferation of product variety occurs at an early stage in the production process. The right side shows a good configuration of the production process in which product variety is determined at a late stage.

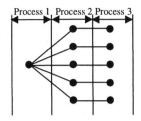
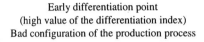
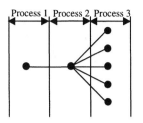

Early differentiation point	Late differentiation point
(high value of the differentiation index)	(low value of the differentiation index)
Bad configuration of the production process	Good configuration of the production process

Figure 9-4. Early versus late differentiation point

In addition to the position of the differentiation point, setup times are decisive for a decrease of work-in-process inventory levels and an improvement of delivery times. Pine (1993, pp. 50) points out that the reduction of setups and changeover costs moves the optimal batch size towards one. In mass customization, ideal setup times are zero, which ensures a high flexibility. Changeovers can be evaluated by the setup cost index (SI) proposed by Martin/Ishii (1996, p. 6). This metric aggregates all setups required for the product and normalizes them with the total costs of all products.

[17]
$$SI = \frac{\sum_{i=1}^{n} v_i\, c_i}{\sum_{j=1}^{v_n} C_j} \; ;$$

$0 < SI < 1$

v_i Number of different products exciting process i

n Number of processes

v_n Final number of varieties offered

c_i Cost of setup at process i

C_j Total cost (material, labour, and overhead) of jth product

(Source: Martin/Ishii 1996, p. 6)

An accurate computation of setup costs may be difficult. Furthermore, when dealing with setups, it is more advantageous to consider the time dimension than the cost dimension. In effect, the capabilities of the process to produce high variety in small batches and to provide short delivery times all depend on time considerations. In opposition to costs, the time dimension provides better indications about the process flexibility level. Therefore, the setup cost index of Martin/Ishii (1996, p. 6) is adapted by replacing costs with lead times and setup durations. Subsequently, we obtain the setup metric *SM*.

[18]
$$SM = \frac{\sum_{i=1}^{n} v_i \, t_i}{\sum_{j=1}^{v_n} T_j};$$

$0 < SM < 1$

v_i	Number of different products exciting process i
n	Number of processes
v_n	Final number of varieties offered
t_i	Average time needed for a setup at process i
T_j	Average total lead time needed for the manfacturing of jth product

For the computation of manufacturing lead times, the product modularity requirement in mass customization must be considered. Modular product architectures involve modular organization of manufacturing on the shop floor. Due to an accurate specification of interfaces, each module can be independently produced in a manufacturing cell (group technology). Therefore, it is legitimate to suppose that the assembly of each module occurs concurrently with the others. Then, all of the modules are delivered to the main assembly line, where they are entirely assembled into a complete product. The total lead time value L is the sum of the time needed for parts assembly into modules, the time for functional testing of modules and the time for modules assembly into the end product (Ericsson/Erixon 1999, pp. 36).

[19]
$$L = \frac{N_p \, T_a}{N_m} + T_t + \frac{N_m - 1}{T_i}$$

L	Lead time
N_p	Number of parts in a complete product
N_m	Number of modules in one average product variant
T_a	Average assembly time for one part
T_t	Average time for functional testing of modules
T_i	Average assembly time for interfaces between modules

(Source: Ericsson/Erixon 1999, p. 118)

The metric L does not take into account the modules delivered by suppliers that do not require assembly in the plant. Furthermore, L considers only one non-value adding activity during manufacturing, namely the functional testing of modules. However, on the shop floor there are many other non-value adding activities such as move and wait times. The adapted key metric T_j for a product j by considering all non-value adding activities, is defined as follows:

[20]

$$T_j = \frac{N_p T_a}{N_m} + T_{nva} + \frac{(N_m + N_s) - 1}{T_i}$$

T_j	Lead time for the manufacturing of product j
N_p	Number of all parts assembled in modules in the plant
N_m	Number of manufactured modules in one average product variant
N_s	Number of supplied modules in one average product variant
T_a	Average assembly time for one part
T_{nva}	Average time for non value added activities
T_i	Average assembly time for interfaces between modules

[21] T_{nva} = move time + wait time + inspection time + setup time

(Source: Maskell 1991, p. 258)

High product variety in mass customization involves steadily changing bottlenecks in production, which induces longer manufacturing equipment idle times that have a bad influence on capacity utilization. Hildebrand/Mertens (1992, p. 71) define the capacity utilization as the ratio of the output to the actually available capacity. Mueller (2001, p. 73) defines a hierarchical system for capacity controlling and suggests evaluating the capacity utilization on the basis of two main parameters namely processing and idle times. Mueller's definition is more precise than Hildebrand/Mertens' definition and will therefore be adopted.

[22]

$$CUM = \frac{\text{Processing time}}{\text{Processing time} + \text{idle time}}$$

CUM Capacity utilization metric

(Source: Mueller 2001, p. 73)

The capability of manufacturing a large number of product varieties with a few production processes and low setup times can be considered as a flexibility measurement in mass customization. Maskell (1991, p. 181) notes that "[t]he fewer processes involved in the production of a product and in the production of the entire plant, the more flexible the plant can be to customer needs. There are two levels of reports that are of value: (1) a report that shows the total number of different processes ... and (2) a report that shows the commonality of these processes across the products currently available for production or currently on the master schedule." However, Maskell (1991, p. 181) does not propose a metric for computing the process commonality level. He recommends deducing commonality by examining the production routings of the finished products and subassemblies. In addition, each company should use a simple definition for commonality that suits its manufacturing processes.

Unlike Maskell (1991), Jiao/Tseng (2000, pp. 232) develop an analytical metric called the process commonality index, which involves many relevant concerns such as process flexibility, lot sizing and scheduling sequencing. This metric is provided by formula [23].

$$
[23] \quad CI^{(p)} = \sum_{p=1}^{n_d} \left\{ \frac{\sum_{j=1}^{n_d} \lambda_{pj} \sum_{j^*=1}^{n^*_{pd}} SET^*_{pj^*} \sum_{j^*=1}^{n^*_{pd}} D_{j^*}}{n_p \sum_{j^*=1}^{n^*_{pd}} \left(D_{j^*} \cdot SET^*_{pj^*} \right)} \left[1 - \frac{n_{pd}}{\sum_{j=1}^{n_{pd}} SET_{pj}} \right. \right.
$$

$$
\left. \left. \sqrt{ \frac{1}{n_{pd}-1} \sum_{j=1}^{n_{pd}} \left[SET_{pj} - \frac{\sum_{j=1}^{n_{pd}} \lambda_{pj} SET_{pj}}{\sum_{j=1}^{n_{pd}} \lambda_{pj}} \right]^2 } \right] \right\} \quad 1 \le CI^{(p)} \le \beta
$$

(Source: Jiao/Tseng 2000, p. 235)

$CI^{(p)}$ Process commonality index

n_p The total number of processes needed for fabrication and assembly

p The index of a particular process, either for fabrication or assembly ($p = 1,2,\ldots,n_p$)

n_d Total number of internally made items, either a final product or component part

j The index of an internally made item, either a final product or component part ($j = 1,2,\ldots,n_d$)

λ_{pj} 0 - 1 variable / $\lambda_{pj} = 1$ if item j can be made or assembled by process p and $\lambda_{pj} = 0$ otherwise

n_{pd}^* The total number of part items produced by process p when production is scheduled in the sequence that minimizes the total setup times ($\sum_{p=1}^{n_p} n_{pd}^* = n_d$)

n_{pd} Total number of of part items required by a product family to be produced by process p regardless of scheduling sequences

j^* The index of an internally made item on process p according to the sequence that minimizes the total setup times ($j^* = 1,\ldots n_{pd}^*$)

SET_{pj}^* The setup time of process p for fabricating or assembling part item j^* according to the sequence minimizing total setup time

SET_{pj} The setup time of process p for fabricating or assembling part j regardless of the sequence minimizing the total setup time

D_{j^*} The demand volume of part item j^*

β The largest possible value of $CI^{(p)}$ ($\beta = n_d$)

The formula proposed by Jiao/Tseng (2000, p. 235) is suitable for the evaluation of the overall process commonality in the case of a product family. A detailed formal description of this formula will not be provided here. However, we provide some informal explanations that enable a better understanding of the notion of process commonality. At first, it can be retained that the production process commonality increases as the capability of each process of producing more different components or subassemblies increases. Suppose that there is only one process that is able to produce all of the different components then the degree of production process commonality can be expected to be high. However, this should be relativized by considering the timeframes that are necessary for the setups. In fact, the adoption of a few processes would be advantageous, when the changeover times that are required by each process for switching from one component to another are low. Otherwise, the involvement of a few processes may be disadvantageous. Consequently, it is necessary to consider the setup times to change from one part item to another. The relevance of the setup times has to be directly related to the part volume. The setup times for the high volume parts can be crucial and should be more heavily weighted than other setup times (Treleven/Wacker 1987, p. 14). That is why the formula involves a term that considers the demand volume of each part item.

In addition, process commonality takes into account the sequencing flexibility, which reflects the freedom of the scheduler to determine the sequence in which the components can be processed or assembled. The higher the sequencing flexibility, the more common the production processes. In this context, Treleven/Wacker (1987, p. 15) mention that the setup time variance is the most appropriate measure for the sequencing flexibility. If for a specific process all of the setup times that are necessary to change over from one component to another are equal, then the scheduler has complete freedom in the selection of the sequence according to which the components are produced. Otherwise, the scheduler has to plan production so that the total setup times are minimized, which involves less flexibility. Thus, the smaller the value of the setup time variance, the higher the process commonality.

4.5   Complexity Key Metrics for the Logistic Sub-Process

The proliferation of product variety may involve an increase of the work-in-process inventory. High inventories are considered a process waste, and dramatically reduce the efficiency of the logistics process. Therefore, the mass customizer has to track the evolution of the work-in-process inventory in the course of time. A suitable key metric enabling the evaluation of the process efficiency is the work-in-process turnover proposed by Pine (1993, p. 112). This key metric is defined as the ratio of total sales to the value of the work-in-process inventory. Before the introduction of new variants, the mass customizer has to evaluate how the key metric value may change. A lower turnover due to higher variety triggers an increase of inventory costs as well as complexity costs.

[24]
$$WIP = \frac{\text{Total sales}}{\text{Value of the work - in - process inventory}}$$

WIP Work - in - process turnover

(Source: Pine 1993, p. 112)

A relevant task of logistics is to deliver the right product at the right point in time. Sometimes the delivery time itself can be the individualization criterion in mass customization. In variety-rich environments, the logistics process has to face a greater challenge than in environments with low variety because of high parts and subassemblies variety, suppliers' variety, production planning and scheduling problems, etc. Many problems can be overcome if the entire supply chain is coordinated in an optimal way. Delivering the product on time to the customer is an indicator that the complete system, including the mass customizer, its partners, and suppliers, at all steps of the value chain work together efficiently. Therefore, we argue that it is important to keep track of the key metric called delivery time reliability.

[25]
$$DR = \frac{\text{Agreed delivery time}}{\text{Real delivery time}}$$

DR Delivery time reliability

4.6   Complexity Key Metrics for the Information Sub-Process

The interaction system is one of the few information systems with which customers directly interact. It consists of two main components, which are the advisory system and the product configuration system. Whereas the knowledge required for advisory can be easily updated and maintained by product experts through a graphical knowledge acquisition component, the knowledge base of the configuration system is more laborious to maintain, especially when the configuration system works on the basis of a rule-based approach.

In order to remain competitive and to respond to changing customers' requirements, the mass customizer may introduce frequent changes at the production program level by adding and/or eliminating product variants. Consequently, the configuration system must be updated as fast as possible.

If the configuration system is not updated in a timely manner, problematical situations can arise, in which the customer orders a product variant that is no longer available in the product assortment. Production program changes may also induce modifications in the way components and product modules interact with each other, which affects the configuration logic.

The continuous assessment of the configuration system performance can be ensured by the key metrics presented by formulae [26] and [27], which are respectively the frequency of introducing changes to the configuration system at a period of time and the average time required to carry out one change. The first metric captures the frequency of updates due to changes in the production program and the second detects the average time required for one update. When these updates are frequent and time consuming, it may be better to replace the configuration system with a more sophisticated one. An other alternative is to increase its integration level in other information systems. However, such investments decisions should be carefully examined and supported by adequate calculations. For instance, one has to compare between the maintenance costs that are triggered by the current configuration system and the costs that are incurred due the implementation of a new system with easier and faster maintenance.

$$FIC\,(\Delta T) = \frac{NC\,(\Delta T)}{\Delta T}$$

[26]

$FIC\,(\Delta T)$		Frequency of introducing changes to the configuration system at a period ΔT
$NC\,(\Delta T)$		Number of changes and database up dates at a period ΔT
ΔT		Period of time

$$AT_c = \frac{\sum_{i=1}^{nc} T_{c_i}}{nc}$$

[27]

AT_c	Average time for carrying out one change in the product configuration system
T_{c_i}	Time required for change i
nc	Total number of changes introduced in the configuration system

In order to reduce the frequency as well as the time required for updating data in the configuration system knowledge base, one can increase the extent to which the configuration system is integrated in the existing information systems, e.g. the Computer Integrated Manufacturing (CIM) systems. The goal of integration is to enhance the automation level. The fewer breaks that exist in the information flow, the more automated the information system will be. As a result, manual interventions required for updating data in different systems are no longer necessary (Rogoll/Piller 2002, pp. 38). For example, when the configuration system is integrated within the Product Data Management (PDM) system, it will suffice to introduce changes in one system, e.g. the PDM system. Subsequently, the database of the configuration system will be automatically updated without manual intervention.

The integration of the configuration system within e.g. the Computer Aided Planning (CAP) system increases the automation degree of order processing in mass customization. Thus, after the customer orders a product variant, the necessary documents for manufacturing such as process routings, assembly instructions, CNC-programs, etc. can be automatically generated in a fast-paced manner. Furthermore, the integration with the Enterprise Requirement Planning (ERP) system enables automatic production scheduling of customer orders as well as automatic materials requirement planning (Blecker/Graf 2003, pp. 5) .

Due to the importance of the integration level of the configuration system within the different components of CIM, we define two key metrics [28] and [29] correspondingly named the integration level of the product configuration system in the existing information systems and the average time elapsed from configuration completion until final preparation of all documents necessary for manufacturing such as routings and task schedules. The key metric [29] captures the speed of the information process in a mass customizing system. The analysis of this metric with respect to different product variants enables the comparison of the times needed by each variant for the preparation of its specific documents. Some product variants may take longer times due to higher technical complexity until the required documents are prepared.

$$IL = \frac{NIP}{NP}$$

[28]

IL	Integration level of the product configuration system within the existing information systems
NIP	Number of information systems integrated within the configuration system
NP	Number of all information systems

$$AT_{(cc \to dp)} = \frac{\sum_{i=1}^{no} T_{(cc \to dp)_i}}{no}$$

[29]

$AT_{(cc \to dp)}$	Average time elapsed from configuration until documents preparation for manufacturing
$T_{(cc \to dp)_i}$	Time elapsed from the completion of configuration i until documents preparation
no	Number of all orders

4.7   Key Metrics Systems for Mass Customization

Since single key metrics have a limited informational value, it is advantageous to integrate them into one or more systems with a stronger explanatory power. In order to ensure the coherence of the key metrics among each other, we propose to evaluate their compatibility with respect to the time pattern that can either be short-term, middle-term or long-term. Figure 9-5 summarizes the retained key metrics as well as the time pattern analysis.

	Key metric	Formula number	Time pattern		
			Short-term	Middle-term	Long-term
Development sub-process	Component commonality	[7]	✓	✓	✓
	Platform efficiency	[9]		✓	✓
	Multiple use	[10]		✓	✓
	Interface complexity	[11]		✓	✓
Interaction sub-process	Used variety	[12]	✓	✓	✓
	Average interaction time	[13]	✓	✓	
	Abortion rate	[14]	✓	✓	
Purchasing sub-process	Purchasing process commonality	[15]	✓	✓	✓
	Differentiation point position	[16]	✓	✓	✓
Production sub-process	Setups	[18]	✓	✓	✓
	Manufacturing lead time	[20]	✓	✓	✓
	Capacity utilization	[22]	✓	✓	✓
	Production process commonality	[23]	✓	✓	✓
Logistics sub-process	Work-in-process inventories	[24]	✓	✓	✓
	Delivery time reliability	[25]	✓	✓	✓
Information sub-process	Frequency of changes	[26]		✓	✓
	Average time needed for changes	[27]		✓	✓
	Integration level	[28]			✓
	Speed of documents preparation	[29]		✓	✓

Figure 9-5. Retained key metrics and time pattern evaluation

The analysis of figure 9-5 reveals that seven key metrics should have values that do not significantly change in the short-term. Among these key metrics, three measures have been identified within the scope of the development sub-process, namely the platform efficiency, multiple use and interface complexity. The other four key metrics are identified during the information sub-process analysis. They are: the frequency of changes, average time needed for changes, integration level and speed of documents preparation. A continuous tracking of these key metrics within small time intervals does not make sense because the corresponding values are expected to significantly change in the middle- and/or long-term.

The platform efficiency measure is the quotient of the average time required to develop a derivative product and the average time to develop a product platform. Product platforms involve high amounts of development costs and time, and are therefore the basic module of a large number of product variants that are launched during a relatively long period of time. In addition, the introduction of derivative products is not a daily, but a strategic decision, which is made on the basis of a middle- to long-term analysis. Subsequently, platform efficiency is a key metric whose evaluation makes sense within relatively large time intervals.

Moreover, the values of multiple use and interface complexity metrics are expected to be constant in the short-term. The key metric values primarily change because of the introduction or elimination of module variants, which rather represents a strategic than a tactical decision.

The information sub-process key metrics are influenced by such investments that aim at improving the integration level of the information systems. Investments are strategic decisions that are generally made on the basis of a long-term analysis. For this reason, it is legitimate to avoid repeatedly computing these key metrics on a short-term basis.

Excepting these key metrics, the other measures that we call short-term key metrics are suitable for the support of tactical decisions and the evaluation of complexity within the short term. It is noteworthy that the so-called short-term metrics are also appropriate for the aggregation of performance values over a middle or long period of time. They are called short-term metrics just to ensure an unambiguous distinction from the key metrics that are exclusively adapted for the middle-/long-term. Thus, a priori the short-term key metrics can be incorporated into a system without violating the time pattern analysis. Nevertheless, although the platform efficiency, multiple use and interface complexity are middle-/long term key metrics, we propose to integrate them into the same system due to obvious correlations with the short-term measures regarding the variety steering problem. This can be further justified by the fact that these three key metrics are identified during the analysis of the development sub-process, which directly interacts with variety as it can be deduced from figure 9-2.

In contrast, the mutual interactions of the information sub-process with variety are not straightforward. As stated previously, the information sub-process is a cross-sectional process that should ensure a smooth information flow between the other sub-processes. In addition, the correlation between the key metrics of the information sub-process and the other measures is not obvious. Moreover, the information sub-process key metrics have to be tracked on a middle-/long-term basis. In order to avoid incompatibilities, it is advantageous to integrate these key metrics into a second system.

To construct both systems, the correlations between the key metrics have to be examined in terms of cause-effect relationships. The intended key metrics systems should make it evident that influence is exerted by the key metrics on each other. It is important to note that both systems do not guarantee completeness. In other words, the change of a key metric value is not *exclusively* due to the change of the value of one or more influencing key metrics represented in the key metrics system. For example, figure 9-6 shows that key metric A has a direct influence on key metric B. It follows that when the value of A changes, then the value of B necessarily changes. However, the value of B may change without necessarily involving a change of A. This can be ascribed to other key metrics or absolute measures (e.g. C, D) which are not represented in the system but actually have an influence on B.

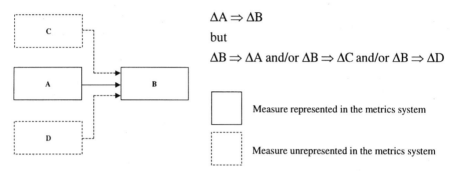

$$\Delta A \Rightarrow \Delta B$$

but

$$\Delta B \Rightarrow \Delta A \text{ and/or } \Delta B \Rightarrow \Delta C \text{ and/or } \Delta B \Rightarrow \Delta D$$

☐ Measure represented in the metrics system

☐ Measure unrepresented in the metrics system

Figure 9-6. Meaning of correlation in the key metrics system

In the following, we propose to gradually construct the first key metrics system by integrating the complexity measures of the development, interaction, purchasing, production and logistics sub-processes into a comprehensive system, which is called the variety-sensitive key metrics system for mass customization. In order to systematically construct this system, it is important to firstly choose a starting point. We argue that it is more legitimate to start from the development and interaction sub-processes. During the development stage, the design engineers define the solution space and the combination possibilities of the product modules, whereas in the interaction phase, the customers select the products that suit their requirements. Both phases can be regarded as a design phase if the customer is considered as a "co-designer" in mass customization.

The complexity key metrics, which are identified within the scope of the development sub-process are: component commonality, multiple use, interface complexity and platform efficiency. Excepting the component commonality that is suitable for supporting tactical decisions in the short-term, the other key metrics are middle-/long-term key metrics, in which to a great extent determine the possible variety that the mass customizer is able to produce. The multiple use metric reflects the flexibility of the product architecture with respect to the possibility to derive a high number of product variants on the basis of a few modules. Furthermore, the interface complexity is an appropriate measure for the evaluation of the modules interface complexity. A high interface complexity may restrain the extent of the possible variety because more tedious coordination between the modules development teams is necessary. In addition, the platform efficiency reflects the ease in deriving new products on the basis of a common platform, which significantly influences the possible variety within the middle-/long-term (Figure 9-7).

It is relevant to note that component commonality is rather determined by the possible variety. Furthermore, it considerably depends on the product variety that is sold to the customer as it can be deduced from the formula [7]. The key metrics that are influenced by component commonality as well as those that influence it are discussed later on.

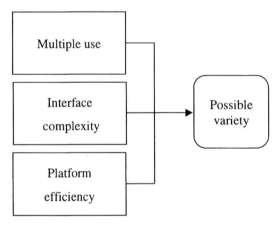

Figure 9-7. Correlations between development key metrics and possible variety

Whereas the key metrics related to the product development influences the possible variety, the interaction key metrics rather affects the perceived variety. Huffman/Kahn (1998, p. 493) have proved that the format in which the product information is presented can strongly influence the variety perceived by the customers. In addition, the customer needs' model which makes the distinction between the objective and the subjective needs suggests that the interaction system plays a relevant role in order to help customers perceive and recognize the optimal product variant. From this it follows that the interaction system has a significant effect on the perceived variety. Note that we have already drawn on two key metrics for the evaluation of the interaction sub-process complexity, namely the average interaction length of time [13] and the abortion rate [14]. It can be concluded that these measures have a direct influence on the perceived variety (Figure 9-8).

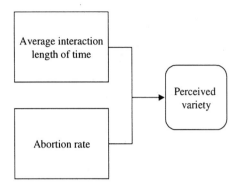

Figure 9-8. Correlations between interaction and perceived variety

Possible variety and perceived variety are not identified as single complexity key metrics. They are respectively the denominator and the numerator of the key metric referred to as used variety. The representations of figure 9-7 and 9-8 emphasize that the middle-/long term key metrics of the development phase influence the possible variety, whereas the key metrics that evaluate the complexity of the interaction sub-process influence the perceived variety. Since the used variety is the quotient of the perceived variety and possible variety, figures 9-7 and 9-8 can be condensed into one representation, which is given by figure 9-9. Thus, we obtain the first part of the key metrics system.

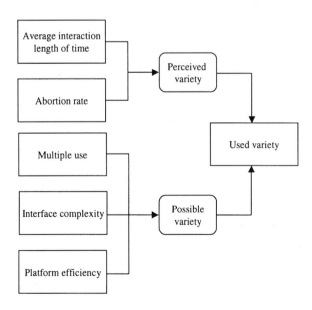

Figure 9-9. First part of the variety-sensitive key metrics system

In order to integrate the rest of the key metrics in the system, we choose the used variety as the new starting point. Among all of the measures, the single key metric that can be influenced by the used variety is the component commonality. Therefore, the influence of this relevant key metric on the other key metrics should be examined.

"While it may be argued that high commonality of internally produced component parts would necessarily mean high commonality of the processes involved, the reverse is not true. Unique items may share similar (or in the extreme case, identical) processes" (Treleven/Wacker 1987, p. 12). As stated by Treleven/Wacker, component commonality has a direct influence on production process commonality. However, production process commonality may be high without involving a high commonality at the component level. This justifies the unidirectional influence of component commonality on production process commonality. When dealing with component commonality, we do not make the distinction between the components that are internally produced and those that are externally outsourced. That is why the overall component commonality may also influence the purchasing process commonality. But the reverse is not true because different components can be outsourced by way of using the same purchasing processes. Figure 9-10 shows the correlations between component commonality and production and purchasing process commonalities.

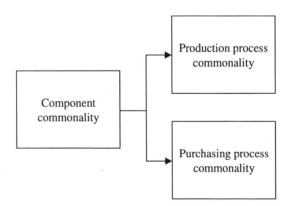

Figure 9-10. Correlations between component commonality and production and purchasing process commonalities

The introduction of a common component that replaces a number of unique components reduces the overall level of safety stock that is required to meet specific service level requirements (Hillier 2000, p. 755). Thus, the component part commonality has a direct impact on work-in-process inventories. Furthermore, common components have a steady demand without fluctuations. Therefore, stochastic methods for the prediction of future demand of these components can be applied with high reliability. Unlike common components, unique components may suffer from high demand fluctuations, which makes a reliable demand prediction difficult and increases the risk of stock-outs.

By increasing the component part commonality level and reducing the number of unique items, the differentiation point (also called decoupling point) can be moved towards the end of the production process. This is justified by the formula of the differentiation index, which takes lower (better) values when fewer component parts and/or subassemblies are involved in the production process. Consequently, component commonality directly influences the position of the differentiation point.

Component commonality also has a major influence on the time required for setup operations. A high component commonality level would mean fewer setups than in the case where the products are manufactured on the basis of a high level of unique components. Figure 9-11 shows the correlation between component commonality, work-in-process inventories, differentiation point and setups.

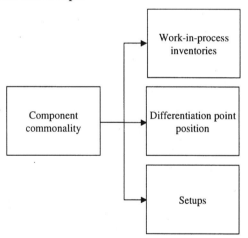

Figure 9-11. Correlations between component commonality, work-in-process inventories, differentiation point and setups

By using the production process commonality index given by formula [23], Jiao/Tseng (2000, p. 238) have examined the impact of setup times on process commonality. When setup times increase, the process commonality index decreases. This result is legitimate because the process commonality index can also be interpreted as a flexibility measure that necessarily changes for the better as the setup times decrease. Conversely, when the setup times decrease, process commonality increases. Furthermore, setups are non-value adding activities that must be taken into account when computing manufacturing lead times. It is common within industry that the value adding activities in manufacturing generally represents a very small part of the total manufacturing lead time. By reducing the non-value adding activities, it is possible to considerably cut the manufacturing lead times, which improves responsiveness and flexibility. The manufacturing lead time is also a term that is required for the computation of the differentiation point position. It is expected that when the lead times are reduced, the differentiation point moves towards the end of the value chain, which considerably improves the delivery reliability of the mass customizer. Thus, we can deduce the impact of manufacturing lead times on the differentiation point position as well as the impact of the differentiation point on delivery reliability. Figure 9-12 shows the correlations existing between setups, manufacturing lead times, differentiation point position and delivery reliability.

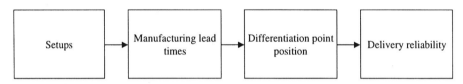

Figure 9-12. Correlations between setups, manufacturing lead times, differentiation position and delivery reliability

By condensing figures 9-10, 9-11 and 9-12 in one representation we obtain the second part of the variety-sensitive key metrics system (Figure 9-13).

By taking into account that the used variety has a direct influence on the component commonality, we can join both parts of the key metrics system in one comprehensive representation (Figure 9-14). It is noteworthy that the capacity utilization metric has been omitted in the variety-sensitive key metrics system for mass customization. It is true that high product variety can induce steadily changing bottlenecks in production, which is associated with longer manufacturing equipment idle times.

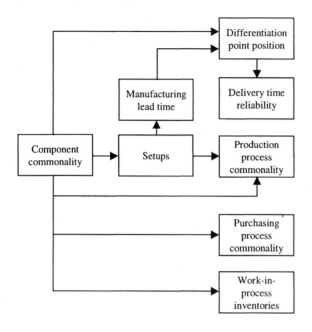

Figure 9-13. Second part of the variety-sensitive key metrics system

However, Fisher/Ittner (1999, p. 785) point out in their study which deals with the impact of product variety on automobile assembly operations that product variety has an insignificant impact on performance once the assembly line has been optimally buffered against the variety induced process time variability with excess capacity. It follows that excess capacity may be even necessary in order to maintain a good performance level. In this case the capacity utilization level seems to not be an appropriate metric to capture the effects of product variety. "Some forms of slack or unused resources may actually improve performance and should not be eliminated in certain manufacturing settings, suggesting that efforts to isolate "excess" capacity costs in activity-based costing systems should be careful not to provide incentives for managers to eliminate all slack resources" (Fisher/Ittner 1999, p. 785). We argue that managers have to track the capacity utilization level but with caution in order to avoid mistaken decisions. In variety-rich environments such as mass customization, excess capacity may be an imperative. Therefore, we have not included this measure in the variety-sensitive key metrics system for mass customization.

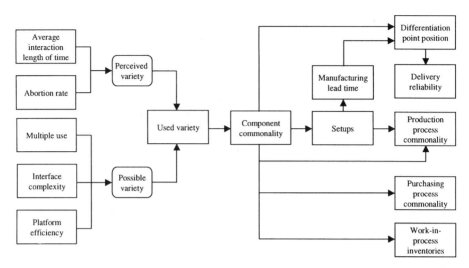

Figure 9-14. Variety–sensitive key metrics system for mass customization

To construct the key metrics system incorporating the complexity measures of the information sub-process, we argue that the key metric related to the integration level of the configuration system in the existing information sub-processes is the starting point. The value of the integration level influences the automation of the preparation of the necessary documents for production. For instance, when the configuration system is integrated with the CAD system, the required drawings of the component parts or subassemblies are automatically generated. A further advantage of increasing the integration level is to avoid data redundancies and inconsistencies. If the configurator is integrated with the PDM-system, any changes in the bill of material introduced by development engineers would automatically initiate updates of the configurator's database. As a result, the integration level has a direct influence on both the average time needed to introduce changes in the product data as well as the frequency of changes. Figure 9-15 presents the key metrics system for the information sub-process in mass customization.

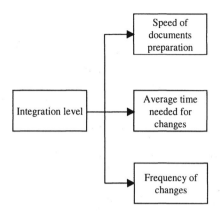

Figure 9-15. Key metrics system for the information sub-process in mass customization

5. EXTENSION OF THE VARIETY-SENSITIVE KEY METRICS SYSTEM FOR MASS CUSTOMIZATION

The variety-sensitive key metrics system has been designed with the objective to detect an increasing or a decreasing of the variety-induced complexity in mass customization. The introduction and/or elimination of product variants are expected to trigger a change of the key metrics values. Accordingly, it can be deduced if complexity has increased or decreased. Due to the importance of the customer's role in mass customization, the decision of introducing or eliminating product variants should to a great extent consider the customer's perspective.

The variety-sensitive key metrics system does not sufficiently take into account the customer's perspective and especially the impact of variety on customers. As stated previously, high product variety can overstrain customers and lead them to select products that do not correspond to their objective needs. It follows that the key metrics system has to be extended in order to measure the extent to which customers are selecting optimal product variants. Note that the customers' objective needs can be viewed from two perspectives. The first perspective that has been already addressed in this book focuses on how to help customers recognize their optimal choice during the interaction process. The second perspective concerns the design of product variants corresponding to the objective needs at the development stage. This last perspective has a critical importance due to its great impact on total performance. If the products are not designed with respect to the objective customers' requirements, then even a good interaction system will not be able to make optimal suggestions to customers. Therefore, we propose an examination of some selected managerial methods that enable the mass customizer to consider the objective needs better at the specification stage of customized products.

In the following, we will deal in more detail with Kansei Engineering that has already been briefly described in chapter five of this book. Kansei engineering technology will be viewed in this section rather as a technology for new product development than as a methodology for the assistance of customers during the interaction process. Then, we outline the key value attributes concept, which aims at determining the product attributes that have to be customized in order to maximize the value to the customer. The relevance of both concepts in dealing with the objective customers' needs is discussed. Furthermore, the difficulties that are related to the identification of the objective needs are highlighted and the influence of variety on customers from a consumer-psychological perspective is shortly characterized. On the basis of the presented ideas, the final key metrics system that should better take into account the impact of variety on customers in mass customization is determined.

5.1   Selected Concepts Supporting the Introduction of Product Variants Fulfilling Objective Customers' Needs

5.1.1   Kansei Engineering Concept

When customers want to buy something, they generally describe their desires in terms of abstract adjectives by using words such as "luxurious", "gorgeous" or "strong", etc. Kansei engineering is technology that enables a consumer's image and feeling (Kansei in Japanese) about the product to be translated into design elements (Nagamachi 1995, p. 4). Kansei engineering is considered as a consumer-oriented methodology for new product development that translates human psychological processes (e.g. emotion and image) into appropriate physical product elements such as size, shape and color. The application of Kansei engineering aims at supporting the customer's decision making as well as the designer's creativity (Yang et al. 1999, p. 460).

There are two types of approaches in Kansei engineering. The first approach is called category classification and aims at sequentially breaking down an abstract design concept of a new product into design elements. Design engineers define the product concept by means of words or sentences as a zero level. Then the zero level concept is broken down to the final stage in which real design elements are precisely defined (Yang et al. 1999, p. 461). Nagamchi (1995, p. 5; 2002, p. 290) reports that Mazda has used this method of category classification to develop a sports car called "Miyata" that has been a good seller in the USA and in Japan. The zero level concept of Miyata made by Mazda is "Human-machine unity". The development project team classifies the zero level concept into sub-concepts at the 1st, 2nd ,..., and n^{th} level until all of the design elements are specified. For example, at the first level the product concept is broken down into four dimensions, which are "Tight-feeling", "Direct-feeling", "Speedy-feeling" and "Communication".

The second approach is the Kansei Engineering System (KES), which is a computer-based decision supporting technique. The KES is composed of three major parts: (1) collection and analysis of Kansei words describing human feelings that are related to the product image, (2) the inference mechanism between human feelings and product design elements, and (3) the presentation of the inference results by means of computer graphics (Yang et al. 1999, p. 461).

From an information technological perspective, the Kansei engineering computerized system, as developed by Nagamachi (1995, pp. 5), can be considered as an expert system that transfers the consumer's feeling and image to the design details. It basically consists of four databases, which are: the Kansei word database, image database, knowledge base and design and color database. In the Kansei word database, the consumer's feelings about a product are stored. The statistical relations between the Kansei words and the design elements are represented in the image database. The knowledge base consists of the rules that are needed to decide upon the highly correlated items of the design details with Kansei words. Finally, the design details are separately implemented in a design database and a color database (Nagamachi 1995, pp. 6-7).

It is important to note that feelings and perceptions have rather an implicit than an explicit character. Therefore, it can be argued that Kansei engineering has a high potential in order to design products according to the objective customers' needs, which are implicit by definition.

5.1.2 Key Value Attributes Concept

Whereas Kansei engineering deals with the design of products with better chances to meet the objective customers' needs, the key value attributes concept enables the mass customizer to optimize variety while maximizing the value to the customer. In this context, we argue that the product variant with the attributes that provide an optimal value to the customer is the variant that fulfills the objective needs. Product attributes can be defined

"…as relatively directly observable physical characteristics of a product or service. Examples are price, colour, weight, etc." (Virens/Hofstede 2000, p. 4). However, customer values are, "…in general defined as relatively stable cognitions and beliefs that are assumed to have a strong motivational impact. Examples are 'security', 'happiness', 'fun and enjoyment', etc." (Virens/Hofstede 2000, p. 4). The personal computer is an ideal example to explain how customer values differ from one person to another. Indeed, a computer is a necessity for individuals who value a sense of accomplishment, status symbol for those who value self-respect and toy for individuals who value fun and enjoyment (Virens/Hofstede 2000, p. 4).

A product variant is obtained by a unique combination of product attributes, whereas the total set of product configurations is generated by the permutations of all attributes. The perceived value of different con-figurations varies from one customer to another. Those attributes with the greatest perceived value to customers are called *Key Value Attributes*. Although product configurations are discrete, the representation of the customer value by means of a curve makes sense due to the high number of product variants that are involved in mass customization (Figure 9-16) (MacCarthy et al. 2002, pp. 76).

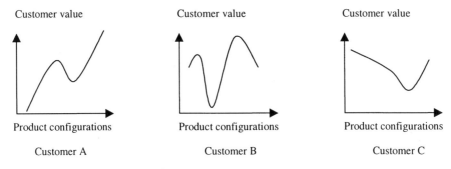

Figure 9-16. Conceptual illustration of customer differences
(Source: MacCarthy et al. 2002, p. 78)

The customer value curves of figure 9-16 are a representation at a defined point in time T. At a later point in time T+ΔT the value curves can change because of changes in many factors such as tastes and fashion, technological innovations and product maturity. The potential for customization can be deduced from differences between customer values, which involves a value difference curve as represented by figure 9-17. A low level of value difference across configurations suggests high market homogeneity, which means that the product can be standardized. However, high difference levels depict high market heterogeneity as well as a greater potential for customization (MacCarthy et al. 2002, pp. 78).

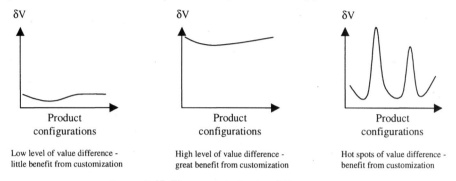

Figure 9-17. The conceptual value difference curve
(Source: MacCarthy et al. 2002, p. 79)

By analyzing the value difference curves, it is possible to determine the product configurations with high difference levels. When we consider the curve on the right side of figure 9-17, the key value attributes are those responsible for high spots. The mass customizer has to recognize these attributes because they represent a high potential for customization. For instance, if customers differently evaluate the attribute 'color', then it is valuable to provide a high color variety from which the customer can find the appropriate one.

The key value attributes concept is relevant for mass customization. It provides interesting approaches for the optimization of product variety according to the objective customers' needs. The product variants with no potential of increasing value to the customers are superfluous and should not be introduced into the product assortment. This concept has an additional relevance because it emphasizes that the attributes to be customized are dynamic and can change over time.

5.2 Problems in Identifying the Objective Customers' Needs

Kansei engineering and the key value attributes concept have been identified to be suitable approaches for the alignment of the product assortment with the objective customers' needs. On the one hand, the Kansei engineering approach provides a product development technique that enables the mass customizer to specify the design elements of the product on the basis of abstract adjectives, which are more suitable for the description of the objective customers' needs. On the other hand, the key value attributes concept aims at maximizing the value to the customer while optimizing product variety. This concept suggests that the attributes that the customers differently evaluate should be customized in order to optimally fulfill the objective customers' needs.

Nevertheless, the use of both methods does not guarantee that the objective customers' needs are perfectly recognized. The identification of the real customers' requirements at the product development phase is difficult and necessitates further research in academia and practice. It is noteworthy that the abilities of traditional methods to capture valuable information about customers' needs are too limited. These methods are unsuitable for understanding the complexity of the consumer's behavior. For instance, an empirical study including small and medium enterprises in Sweden has shown that the involved companies were disappointed by the results of their customer interviews. The companies expected that their customers could express their needs by using product specific terms and making the sometimes implicit, explicit. However, the participants (potential customers) were generally not able to formulate their requirements and did not propose innovative solutions (Ekstroem/Karlsson 2001, p. 24).

In order to better consider the objective customers' needs at the development phase, managers can use the Kano's model as described by figure 9-18. This model makes the distinction between the basic, performance and excitement attributes. The basic attributes are the 'must-be' attributes whose optimal fulfillment is taken for granted, and will only lead to a state of 'not dissatisfied' (Matzler/Hinterhuber 1998, p. 28). However, a poor fulfillment of the basic attributes triggers extreme customer dissatisfaction. For example, the braking system is a basic attribute in a car that is expected to be optimal in performance. If the braking system does not operate in an optimal manner, then customers are extremely dissatisfied.

The performance attributes correspond to the customers' needs that can be articulated. However, it frequently happens that customers do not adequately express the corresponding needs. A suboptimal fulfillment of the performance attributes negatively influences customer satisfaction. For example, customers are less satisfied with a car that does not provide fuel economy. Finally, the excitement attributes are neither articulated nor expected. They represent latent needs of which customers are unaware. The fulfillment of these attributes leads to extreme customer satisfaction and provides the company with a considerable advantage over its competitors. In addition, these attributes should exponentially inspire customers (Berger et al. 1993, pp. 4).

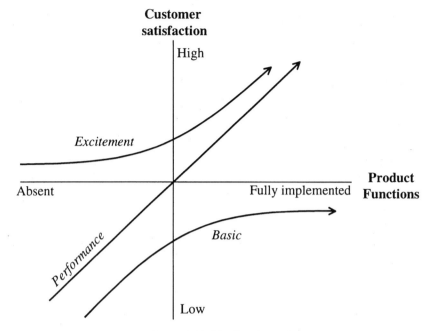

Figure 9-18. The Kano's model

(Source: Berger et al. 1993, p. 4)

According to the Kano's model, the excitement attributes fulfill the customers' needs that are customer tailored, transcendent and not expressed (Matzler/Hinterhuber 1998, p. 28). On the other hand, following the results of chapter four of this book, the objective customers' needs are implicit, unarticulated and lead to optimal customer satisfaction. Thus, with respect to both definitions, obvious correspondences exist between the objective customers' needs and excitement attributes. That is why it can be stated that the excitement attributes play an important role in meeting the objective customers' needs and that the mere fulfillment of the expressed individual customers' needs does not trigger optimal satisfaction. In this context, Matzler/Hinterhuber (1998, p. 28) argue, "...fulfilling individual customer expectations to a great extent does not necessarily imply a high level of customer satisfaction. It is also the type of expectation that defines the perceived product quality and thus customer satisfaction." However, the excitement attributes are the results of innovative and creative ideas of the mass customizer. In order to better align the product assortment with the real requirements of the customers, Ekstroem/Karlsson (2001, p. 24) suggest keeping a permanent dialogue with consumers during product development.

5.3   Impact of Variety on Customers from a Consumer Psychological Perspective

It is obvious that the product variants corresponding to the objective needs can be better perceived by customers in contexts where product variety is not very extensive. This suggests that product variety may have an impact on customers from a consumer psychological perspective. Desmeules (2002, pp. 6) has examined the relationship existing between variety and consumer behavior. He provides a graphical model that describes how variety can correlate with the positiveness of a consumption experience when customers evaluate the product variants by cognition (Figure 9-19). The dependent variable "positiveness of a consumption experience" could either be customer happiness or satisfaction. Whereas customer satisfaction is a post-purchase evaluation of a product or a service, customer happiness extends the meaning of customer satisfaction and includes the shopping experience. In the web-based mass customization, the shopping experience refers to customer interaction via the Internet.

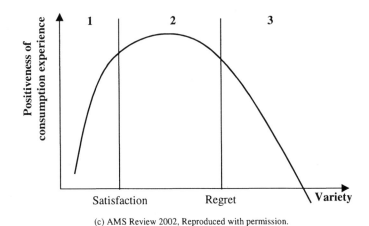

(c) AMS Review 2002, Reproduced with permission.

Figure 9-19. Relationship between perceived variety and positiveness of consumption experiences when the evaluative task is performed by cognition (Source: Desmeules 2002, p. 10)

The inverted "U"-shaped relationship between variety and the positiveness of a consumption experience presents three sections. Section (1) indicates that the addition of new product variants increases customer happiness because the likelihood that customers will find the variant they are looking for is higher. The point of satisfaction is reached when all of the variants that customers require are available. The product variants of section

(2) do not have an influence on the consumption experience and may either be considered or ignored by customers. The end of section (2) refers to the point of regret where customer happiness starts to considerably decrease. In section (3), it is assumed that the addition of product variants negatively affects the consumption experience because of stress, frustration and regret. Regret is experienced because of the high number of product variants that leads to an information overload. Subsequently, consumers would feel that the selected product variant is not optimal and that another product configuration in the assortment would be more suitable for them (Desmeules 2002, p. 10).

Iyengar/Lepper (2000) posit that in limited-choice contexts people are engaged in rational optimization, whereas in extensive-choice contexts people simply give up the choice-making process after they find a product that is merely satisfactory, rather than optimal. In this context, Schwartz (2000, p. 21) argues that the addition of new options may influence the choice situation that may become less rather than more attractive. In order to overcome such situations in which product variety is no longer straightforward, customers generally look for the help of product experts who make the decisions for them.

5.4 Extended Key Metrics System for Mass Customization

The extended key metrics system will integrate all of the concepts and ideas that have already been discussed. It should enable the evaluation of both the complexity that is induced by variety and the extent to which the mass customizer fulfills the objective customers' needs. The extended key metrics system is based on two major parts. The first part is the variety-sensitive key metrics system described by figure 9-14. The second part will be determined in this section and represents the extension that takes into account the customer's perspective. Then, both parts are integrated together.

At the development stage, Kansei engineering provides the mass customizer with valuable support for the design of products according to the objective customers' needs. Therefore, it is recommended to use this technology for new product development in mass customization. On the other hand, in order to ascertain the attributes over which customers are allowed to express choices, the key value attributes concept suggests examining the value difference curves. However, the curve shapes change over time, which suggests that some attributes no longer maximize the value to the customer. For this reason, new key value attributes should be considered as customizable options. MacCarthy et al. (2002, pp. 74) have identified five factors that have a major influence on the shapes of the value difference curves, which are: tastes and fashion, different markets, competitive environment, product technological capability and product innovation, as well as product maturity. Changing customizable attributes has already been observed in practice, e.g. in the automotive industry. Indeed, some options that were offered in the past to be customized are now serially produced. Another relevant approach is to bundle options in packages (e.g. sports package) so that customers do not make choices over single options, but over packages. Packaging and the well-founded reduction of the number of customizable options decrease product variety as well as the variability between end product variants, while retaining high value for the customer. Both approaches have an additional advantage because they increase performance in manufacturing and considerably reduce complexity costs.

By excluding the case of sectional modularity in which the product attributes can be determined by the way the modules are connected (e.g. Sofa U-shape and Sofa S-shape), it is generally true that each product attribute corresponds to a specific product module. Indeed, "...by definition of modular product architecture, there is a one-to-one mapping of product functions, which are the engineering equivalents of product attributes, onto components" (Salvador et al. 2002, p. 569). It follows that the introduction, elimination or bundling of product attributes is equivalent to the introduction, elimination or bundling of physical modules. In addition, the attribute levels correspond to the module variants. If we consider the attribute or module 'exterior color' of a car then the attribute levels or module variants can be e.g. blue, red, silver, white and black.

A continuous updating of the product assortment by introducing and/or eliminating product attributes and/or attribute levels with respect to the key value attributes concept is essential in mass customization. The main goal is to respond to changing objective customers' preferences and desires. In addition, the customizable attributes to a great extent determine the performance of the mass customizer because they considerably influence the internal complexity from the mass customizer's perspective and external complexity from customer's perspective. Therefore, we argue that it is relevant to track the number of customizable attributes [30] as well as the average number of customizable attribute levels [31].

[30] N_{ca} Total number of customizable attributes

$$AN_{al} = \frac{\sum_{i=1}^{N_{ca}} (N_{al})_i}{N_{ca}}$$

[31]

AN_{al} Average number of attribute levels

$(N_{al})_i$ Number of attribute levels per attribute i

N_{ca} Total number of customizable attributes

The impact of the number of customizable options on customers is explained by the model of Desmeules (2002, p. 10) examining the relationship between the extent of the offered variety and customer happiness (customer satisfaction with the shopping experience and with the product). A possible interpretation of this model is that the likelihood a customer selects a product, which corresponds to his/her subjective needs, would increase after the point of regret owing to information overload. Indeed, if customers are overwhelmed by product variety, they tend to cease the searching process when they find a merely satisfactory alternative even when an optimal product variant that better satisfies their requirements may exist in the product assortment. As a result, customer happiness considerably decreases because the objective needs will not be fulfilled. However, around the point of satisfaction, product variety is more straightforward from the customer's perspective. Subsequently, the likelihood that the offered variety matches the objective customers' needs is high, which increases customer happiness.

Thus, it can be stated that so long as product variety improves customer happiness, the probability of matching the objective customers' needs increases. However, if variety worsens customer happiness, then the likelihood of fulfilling the objective customers' needs decreases and the introduction of new product variants will rather be confusing and frustrating for customers. Therefore, customer happiness is a suitable approach for evaluating the extent to which the objective customers' needs are fulfilled. In order to track customer happiness, customer surveys can be used. However, surveys cannot be frequently implemented because of their high costs. They are unsuitable for permanent tracking of customer happiness on a short-term basis. But the level of customer happiness can be deduced by using key metrics such as churn rate, return rate and complaint rate which are sensitive and require data that is commonly available. Note that when the churn, return and complaint rates increase, customer happiness decreases.

$$CCR\,(\Delta T) = \frac{NOLC\,(\Delta T)}{NOC\,(T) + NONC\,(\Delta T) - NOLC\,(\Delta T)}$$

	$CCR(\Delta T)$	Customers churn rate at the period of time ΔT
[32]	$NOLC\,(\Delta T)$	Number of lost customers at the period of time ΔT
	$NOC\,(T)$	Number of customers at the point in time T
	$NONC\,(\Delta T)$	Number of new customers at the period of time ΔT

(Source: Sterne 2002, p. 146)

$$RR\,(\Delta T) = \frac{\text{Number of returned products}\,(\Delta T)}{\text{Number of delivered products}\,(\Delta T)}$$

[33]

$RR(\Delta T)$ Return rate at the period of time ΔT

(Source: Piller 2002, p. 16)

$$CR\,(\Delta T) = \frac{\text{Number of complaints}\,(\Delta T)}{\text{Number of deliveries}\,(\Delta T)}$$

[34]

$CR(\Delta T)$ Complaints rate at the period of time ΔT

With respect to the churn rate metric, the main question relates to the rule to be considered for the determination of migrated customers. In fact, each mass customizer can adopt a specific rule that suits its business. For example, it is conceivable to consider the customers who have not repurchased within a certain period of time as lost customers, which means in this case that the churn rate depends on the frequency of repurchase.

The complaints rate metric has to be addressed with precaution because "...in general only about 5% of unsatisfied customers ever complain" (Walczuch/Hofmaier 1999, p. 7). In order to have an idea about the actual complaints rate, the computed key metric value must be appropriately amplified.

Happy customers do more business and purchase more frequently (Brown/Gulycz 2002, p. 34). Therefore, it is expected that customer happiness positively influences the repurchase rate which is defined as the quotient between the repurchase volume through existing customers in a period of time and the purchase volume through new customers in the same period of time. It is worth noting that the customer will discover that the received product does not correspond to his or her objective needs during consumption. If the product's warranty has not expired then the customer can return the product to the mass customizer, which is captured by the return rate metric. However, if the product's warranty is expired then the customer may complain and the likelihood that he/she repurchases a product from the mass customizer can be expected to be low. Consequently, the repurchase rate is an adequate metric to evaluate the extent to which the objective customers' needs are fulfilled.

$$R(\Delta T) = \frac{\text{Repurchase volume through extisting customers } (\Delta T)}{\text{Purchase volume through new customers } (\Delta t)}$$

[35]

$R(\Delta T)$ Repurchase rate at ΔT

(Source: Piller 2002, p. 15)

There are obvious correlations between the different key metrics that are determined in this section. The model of Desmeules (2002) suggests that the customizable attributes and the average number of attribute levels have a direct influence on customer happiness, which in turn positively influences the repurchase rate. These correlations can be represented by figure 9-20.

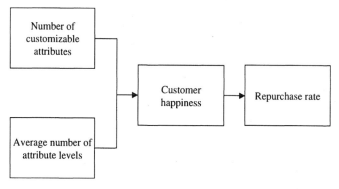

Figure 9-20. Key metrics system taking into account the customer's perspective

As previously mentioned, there is a one-to-one mapping between product attributes and product modules within a modular product architecture, which suggests that some correlations may exist between the measures of the key metrics system of figure 9-20 and those of the variety-sensitive key metrics system. Note that the performance of the modular architecture is captured by measures that have been already specified in the variety-sensitive system, namely "multiple use", "interface complexity" and "platform efficiency". The "multiple use" metric evaluates the flexibility of the product architecture. In addition, it is legitimate to state that the introduction of new customizable attributes is positively driven by the flexibility of the product architecture. Therefore, we expect a positive correlation between the multiple use metric and the number of customizable attributes. However, the interface complexity may represent a constraint for the introduction of customizable options. Indeed, the inflexibility and incompatibility of module interfaces can be an obstacle for the integration of new options into product configurations, which means that interface complexity negatively affects the number of customizable attributes. Furthermore, as aforesaid, platform efficiency assesses the ease of deriving new product variants on the basis of a product platform. If the platform does not already envisage, at its development stage, the introduction of some options during the life cycle of the product, then costly and time-consuming adaptations may be required, which considerably restrains the number of customizable attributes. On the other hand, it is obvious that the number of customizable attributes and attribute levels directly influence the product variety that is perceived by customers. However, it is not possible to definitively claim that this correlation is positive or negative. The limited capacity of human beings to process information and some approaches from consumer psychology allow one expect that the perceived variety may increase with each introduction of a new customizable attribute until a limit value is reached. After this value the increasing number of product options will rather negatively than positively affect the perceived variety.

As the number of customizable attributes and attribute levels increase, component commonality may be negatively affected because the likelihood that new components are required also increases. However, the production process commonality is influenced only by the customizable attributes number. Indeed, different modules may require different processes inside the manufacturing system. But the implementation of group technology in manufacturing (modular manufacturing systems) as a consequence of a modular product architecture suggests that the different levels of one product attribute can be produced on the basis of the same process. That is why we expect no direct influence of the attribute levels on the production process commonality.

By taking into account all of the identified correlations we can derive the extended key metrics system for mass customization which is presented by the following figure 9-21.

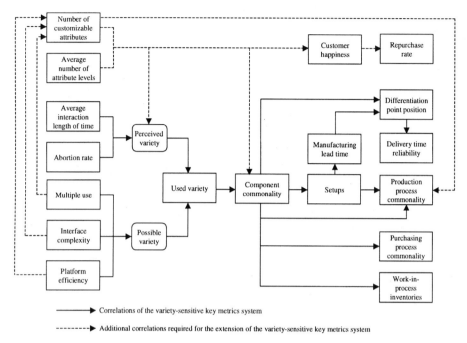

Correlations of the variety-sensitive key metrics system

Additional correlations required for the extension of the variety-sensitive key metrics system

Figure 9-21. Extended key metrics system for mass customization

The extended key metrics system for mass customization has to be applied per product family. It cannot be used to aggregate data of different product families, which are generally different with respect to the product platforms, production processes, customers' base, etc. Furthermore, target values for each key metric have to be defined because key metrics have no relevance, if they cannot be compared with predefined targets. For instance, the mass customizer can benchmark its performance with other mass customizers or make-to-order companies being active in the same or different fields. It is also conceivable to create a website where several mass customizers put their data online in order to ensure continuous key metrics-based benchmarking.

6. SUMMARY

In this chapter, we have pointed out the drawbacks of the current approaches (Pareto analysis, contribution margin accounting method and activity based costing) in dealing with the variety-induced complexity in mass customization. In order to cope with the main disadvantages, we have decided to provide another solution approach based on key metrics systems, which are identified to be suitable management tools.

On the basis of the main internal abilities of the mass customizing system defined in chapter two, we have determined the main sub-processes in mass customization. The mutual interactions between the different sub-processes and variety are analyzed. The sub-process analysis has enabled us to better structure the complexity problem. For each sub-process, the relevant key metrics required for an efficient evaluation of the variety-induced complexity are determined. However, single key metrics do not have strong explanatory power. That is why they should be integrated in a comprehensive key metrics system depicting the different correlations existing between them. Furthermore, it was not possible to integrate all of the identified key metrics in one single system because of some incompatibilities. For this reason, two key metrics systems are established. The first system incorporates the variety-sensitive key metrics, whereas the second system integrates the different measures that are appropriate for the evaluation of complexity in the information sub-process.

The variety-sensitive key metrics system consists of the relevant measures to be used for variety steering. This system can capture the complexity increasing or decreasing effects of variety. Managers can use this key metrics system in order to evaluate the internal complexity that is induced by new product variants. However, the variety-sensitive key metrics system does not take into account the customer's perspective, which is of high relevance in mass customization. Therefore, the idea was to further extend the system with key metrics that are able to evaluate the extent to which the mass customizer fulfils the objective customers' needs. The extended key metrics system provides valuable information about the internal complexity inside production and manufacturing-related tasks. Furthermore, it enables one to measure the extent of the external complexity, which by definition hinders customers to select products that correspond to their objective needs.

Chapter 10

CONCLUSIONS

Customization is a continuously growing business trend that aims at providing customers with individualized goods and services. In dynamic business environments, it is even a necessary condition in order to maintain competitive advantage and outpace competitors. The discussion presented in this book concentrates on the challenging case, in which manufacturing enterprises strive for customizing physical goods by taking into account the costs' efficiency perspective. Mass customization is an oxymoron that joins both perspectives of product customization and mass production efficiency into one concept. In order to lead mass customization to success, there are some necessary conditions to be satisfied. A distinction should be essentially made between the conditions to be satisfied before shifting to mass customization and those to be maintained and further developed during the pursuit of the strategy.

Mass customization induces a high level of product variety, which increases the internal complexity in operations and manufacturing related tasks. The effects of complexity generally arise in the form of hidden costs that trigger efficiency problems. Mass customization involves high production program complexity, high configuration complexity for customers and increasing planning and scheduling complexity. However, mass customization not only increases complexity, but it has some potential for reducing complexity. The pursuit of mass customization drives the reduction of complexity at three main levels, which are: the order taking process, product and inventory. The success of mass customization can only be reached if the complexity increasing factors are adequately managed. Another important issue for the achievement of a successful mass customization is an optimal understanding of customers who should be considered as partners in the value creation process. In the specific context of mass customization, customer needs are frequently misunderstood. It frequently occurs that customers find out that the selected product configuration does not exactly meet their expectations right after delivery or during consumption. The model making the distinction between the subjective and objective customer needs reveals many insights about the nature of customer requirements in mass customization. The subjective customers' needs are the individually realized and articulated requirements, whereas the objective needs are the real ones perceived by a fictive neutral perspective. The subjective needs are explicit and lead to sub-optimal customer satisfaction. But the objective needs are implicit and yield optimal customer satisfaction. The customer needs' model provides interesting clues for making optimal decisions concerning the variety steering problem. It suggests the elimination of the over engineered variants that neither correspond to the subjective nor to the objective needs. Another important result is that the product variants that only correspond to the subjective customers' needs have to be eliminated. These product variants are especially problematical because they confuse customers. Since customers identify their subjective needs, they may believe that a product alternative is optimal. However, there is another product variant in the solution space that fulfills the objective requirements. The result is that customers select product variants that do not exactly meet their requirements. In addition, the model suggests that customers can better recognize their objective needs if the external complexity faced during the product configuration process is considerably reduced.

The customers' needs model for mass customization reveals that customers have to be adequately supported during the product selection process in order to make them identify their objective needs. In the web-based mass customization, the customer assistance task is supposed to be ensured by online configuration systems. These software tools enable the customer-supplier interaction and should play an important role in the customer needs' elicitation process. However, the examination of the morphological box that includes the main dimensions according to which the configuration systems are developed, suggests that the state of the art configurators are very product oriented. They assume that customers are able to make rational decisions in extensive variety contexts, which is not true because large variety confuses customers. In order to mitigate this main drawback, the extension of configuration systems with an advisory system is necessary. This software system takes over the relevant task of capturing the objective needs and assisting customers during the decision-making process. For the development of the advisory system, two main steps were necessary. The result of the first step is a software system that is based on a hybrid approach including a knowledge base and recommender techniques. This basic system takes over the task of human advisors by asking the customers suitable questions in order to capture their preferences and needs. Then, the advisory system proposes a limited number of product suggestions, from which the customer can choose the appropriate alternative. The second step of the development process deals with the improvement of the system performance with respect to its capability of optimally eliciting the objective customers' needs. The logic, according to which the system works, remains the same, which means on the basis of customer dialogs in order to make suitable product suggestions. However, the final implementation includes additional technologies such as a CRM system, web mining technology, and Kansei engineering. These technologies are appropriate solutions for an optimal support of the objective needs' elicitation. The advisory system only enables the mitigation of the external complexity problem and does not cope with the internal complexity. A comprehensive software system based on agent technology enables one to additionally address the internal complexity problem. For the development of this system, the products to be customized are assumed to be manufactured on the basis of modular and platform strategies. To each module variant and platform, autonomous rational agents are assigned. In order to work effectively, it is necessary to integrate the module and platform agents into a framework that additionally contains a configuration system, advisory system, target costing agent, auction agent, product constraints' agent and validation agent. The advisory system captures customer preferences and maps them to product functionalities. Then the target costing agent determines the product platforms and modules that could form product variants, in which the customer would be interested. The selected module and platform agents negotiate with each other in order to form consistent product variants. The Dutch auction is the most suitable mechanism that supports the reaching agreement process. That is why the corresponding agent is called the Dutch auction agent. It determines the auction functions and communicates them to the platform agents who initiate the variety formation process. Because the product constraints' agent has direct access to the product logic of the configuration system, it only allows the formation of consistent product variants. Finally, the validation agent selects the product variants that best correspond to customer requirements. These variants that are displayed to the customer are considered to be successful. A reward mechanism recompenses the module agents that participate in the formation of successful product variants. The notion of a module agent's account enables the implementation of the reward mechanism. Each module agent has an account of money that constantly decreases in the course of time. If the module agent is successful, then it can earn money and extend its life cycle. However, if the module agent does not participate in the formation of successful product variants, then it loses money. When the module agent's account is exhausted, the module agent has to be considered for eventual elimination. By tracking the agents' accounts, managers can evaluate the success of the modules and their suitability in fulfilling customer's requirements. It is important to emphasize that whereas the implementation of an advisory system is appropriate when the external complexity is high and the internal complexity is low, the multi-agent based system is an adequate solution when both external and internal complexities are high. The multi-agent based approach has a higher complexity because it has to retrieve data from several autonomous data sources that must be interfaced.

For the development of the multi-agent based approach, the modularity of products is an important assumption. Product modularity is a relevant requirement in order to put the "mass" in mass customization. The development of products around a modular design has many managerial and organizational implications to be considered by the mass customizing enterprise. For instance, product modularity has major influences on the intra-firm and inter-firm levels. Though modularity enables the reduction of the costs of variety, the complexity costs that are induced by variety can be only brought down if the product assortment is based on a few modules. But variety in mass customization can be very large to where a high number of product modules may be required. In order to control the variety-induced complexity in mass customization, key metrics systems including measures that can be calculated on the basis of available data in the company, are appropriate managerial tools. The sub-process analysis is a structured method for the identification of the relevant key metrics. The integration of these key metrics into a comprehensive system is advantageous because key metrics systems have a stronger explanatory power than single measures. A time pattern analysis that aims at identifying the long term, middle term and short term key metrics reveals that the establishment of one single key metrics system is impossible because of some incompatibilities. The integration of these metrics into two different systems is necessary in order to avoid the incompatibility problems. The first system incorporates the variety-sensitive key metrics, whereas the second system integrates the measures that are appropriate for the evaluation of complexity in the information sub-process. The variety-sensitive key metrics system consists of the relevant measures to be used for variety steering. It captures the complexity increasing effects of variety. Managers can use this key metrics system for the evaluation of the internal complexity that is induced by the introduction of new product variants. However, the variety-sensitive key metrics system does not consider the customer's perspective. An extension of this system with the appropriate measures that evaluate to which extent the mass customizer meets the objective customers' needs is fundamental.

Consequently, the extended final key metrics system assesses the internal complexity inside production and manufacturing-related tasks as well as the extent of the external complexity.

Making mass customization work efficiently is certainly not an easy task. Throughout the discussion presented in this book, managerial and information technological tools are elaborated in order to adequately cope with the main problems that are encountered while pursuing the mass customization strategy. However, this research has opened many interesting questions, which can be considered as directions for future research. The following is a list of research topics that could be of interest:

1. The investigation of the necessary conditions for the successful pursuit of mass customization – or as many researchers prefer to refer to as "success factors" – provided in this book are based on comprehensive literature research. It would be interesting to empirically validate the identified factors through surveys including mass customizers from many industrial fields.

2. The customer needs' model that makes the distinction between the objective and subjective customers' needs reveals many important research issues. The advisory system can be considered as a first step towards a comprehensive solution for the problem of providing customers with products that exactly correspond to their real needs. The identification of the objective needs is challenging, especially when transactions are carried out over the Internet. It is a multidisciplinary problem that necessitates contributions and competences from many fields such as business administration, artificial intelligence, computational technology, and consumer psychology. Thus, sophisticated methods are necessary in order to understand the real customers' needs, through e.g. the exploitation of the customer's navigation behavior.

3. The interaction systems in mass customization to a great extent determine the success or failure of the whole customization process. The development of adequate tools (e.g. based on key metrics) or check lists for the evaluation of the performance of these software systems could be helpful for managers. In effect, with such tools, it is possible to assess the performance of an existing system with the objective to determine enhancement potentials or to compare between different systems before deciding which one to be developed or implemented.

4. Up to now, there are only a few methods for a systematic development of modular product architectures. Therefore, further work is necessary in this field. Another important issue is the development of techniques for the assessment of the optimality of a modular concept.

5. In order to assess the variety induced complexity in mass customization, it is pointed out that key metrics are suitable approaches. An interesting issue could be to develop key metrics for the evaluation of e.g. the entire product assortment or manufacturing system complexity in mass customization. In this context, a good candidate may be the entropic measure. For instance, it can be analyzed as to how the manufacturing system or product assortment entropy changes when new product variants are introduced in the production program.

6. The key metrics systems that are provided in chapter nine show the different correlations existing between the selected measures. However, it would be advantageous to find the mathematical relationships between the main measures, such as e.g. the differentiation index as a function of component commonality or the process commonality as a function of component commonality, etc. so that it would then be possible to quantify the influence of each key metric on another.

Summarizing, one could say that the concept of mass customization is promising but challenging at the same time. It introduces a new business era, in which each person can receive an individualized product at an affordable price. Pine has mentioned in his introduction to the 2nd interdisciplinary World Congress on Mass Customization and Personalization that as he wrote his book in 1993, mass customization was the new frontier. However, today, mass customization is the new imperative for businesses. Thus, mass customization is not a buzzword but a practice relevant competitive strategy. In order to make mass customization "...as important in the 21st century as Mass Production was in the 20th century" (Pine 2003), mass customization must be further supported by academia and practice. Innovative ideas are required in order to strengthen the fundaments for a successful pursuit of this business paradigm.

References

Adam, Nabil R. / Atluri, Vijayalakshmi / Adiwijaya, Igg (2000): SI in Digital Libraries, Communications of the ACM, Vol. 43, No. 6, pp. 64-72.

Anderson, David (1997): Agile Product Development For Mass Customization, Chicago-London-Singapore: IRWIN Professional Publishing 1997.

Anderson, David (2004): Build-to-order & Mass Customization – The Ultimate Supply Chain Management and Lean Manufacturing Strategy for Low-Cost On-Demand Production without Forecasts or Inventory, Cambria, California: CIM Press 2004.

Ardissono, Liliana / Felfernig, Alexander / Friedrich, Gerhard / Jannach, Dietmar / Zanker, Markus / Schaefer, Ralph (2003): A framework for the development of personalized, distributed web-based configuration systems, AI Magazine, Vol. 24, No. 3, pp. 97-110.

Ashby, Ross, W. (1957): An Introduction to cybernetics, 2nd Edition, London: Chapman & Hall LTD 1957.

Baldwin, Carliss Y. / Clark, Kim B. (2000): Design Rules – The Power of Modularity: Cambridge / Massachusetts: The MIT Press 2000.

Baldwin, Carliss Y. / Clark, Kim B. (2003a): Managing in an Age of Modularity, in: Raghu Garud / Arun Kumaraswamy / Richard N. Langlois (Eds.): Managing in the Modular Age – Architectures, Networks, and Organizations, Malden et al.: Blackwell Publishing 2003, pp. 149-171.

Baldwin, Carliss Y. / Clark, Kim B. (2003b): Where Do Transactions Come From? A Perspective from Engineering Design, Working Paper, Harvard Business School, URL: http://www.people.hbs.edu/cbaldwin/DR2/TransactionsFeb5v4.pdf (Retrieval: July 15, 2004).

Barker, Virginia E. / O'Connor, Dennis E. (1989): Expert Systems for Configuration at Digital: Xcon and Beyond, Communications of the ACM, Vol. 32, No. 3, pp. 298-318.

Benameur, Houssein / Chaib-draa, Brahim / Kropf, Peter (2002): Multi-item auctions for automatic negotiation, Information and Software Technology, Vol. 44, No. 5, pp. 291-301.

Berger, Charles / Blauth, Robert / Boger, David / Bolster, Christopher / Burchill, Gary / DuMouchel, William / Pouliot, Fred / Richter, Reinhart / Rubinoff, Allan / Shen, Diane /Timko, Mike / Walden, David (1993): Kano's Methods for Understanding Customer-defined Quality, Center for Quality Management Journal, Vol. 2, No. 4, pp. 3-36.

Berman, Barry (2002): Should your firm adopt a mass customization strategy?, Business Horizons, Vol. 45, No. 4, pp. 51-60.

Blecker, Thorsten / Abdelkafi, Nizar / Kaluza, Bernd / Friedrich, Gerhard (2003a): Variety Steering Concept for Mass Customization, Working Paper No. 2003/04, University of Klagenfurt, URL: http://wiwi.uni-klu.ac.at/2003_04.pdf (Retrieval: July 15, 2004).

Blecker, Thorsten / Abdelkafi, Nizar / Kaluza, Bernd / Friedrich, Gerhard (2003b): Key Metrics System for Variety Steering in Mass Customization, in: Frank T. Piller / Ralf Reichwald / Mitchell M. Tseng (Eds.): Competitive Advantage Through Customer Interaction: Leading Mass Customization and Personalization from the Emerging State to a Mainstream Business Model, Proceedings of the 2nd Interdisciplinary World Congress on Mass Customization and Personalization - MCPC'03, Munich, October 6-8, 2003, pp. 1-27.

Blecker, Thorsten / Abdelkafi, Nizar / Kreutler, Gerold / Friedrich, Gerhard (2004a): An Advisory System for Customers' Objective Needs Elicitation in Mass Customization, Proceedings of the 4th Workshop on Information Systems for Mass Customization (ISMC 2004) at the fourth International ICSC Symposium on Engineering of Intelligent Systems (EIS 2004), University of Madeira, Funchal / Portugal, February 28 - March 3, 2004, pp. 1-10.

Blecker, Thorsten / Abdelkafi, Nizar / Kreutler, Gerold / Friedrich, Gerhard (2004b): Dynamic Multi-Agent Based Variety Formation and Steering in Mass Customization, in: Isabel Seruca / Joaquim Filipe / Slimane Hammoudi / José Cordeiro (Eds.): Proceedings of the 6th International Conference on Enterprise Information Systems (ICEIS 2004), Porto / Portugal, April 14-17, 2004, Volume 2, pp. 3-13.

Blecker, Thorsten / Abdelkafi, Nizar / Kreutler, Gerold / Friedrich, Gerhard (2004c): Product Configuration Systems: State of the Art, Conceptualization and Extensions, in: Abdelmajid Ben Hamadou / Faiez Gargouri / Mohamed Jmaiel (Eds.): Génie logiciel et Intelligence artificielle. Eight Maghrebian Conference on Software Engineering and Artificial Intelligence (MCSEAI 2004), Sousse / Tunisia, May 9-12, 2004, Tunis: Centre de Publication Universitaire 2004, pp. 25-36.

Blecker, Thorsten / Abdelkafi, Nizar / Kaluza, Bernd / Kreutler, Gerold (2004d): Mass Customization vs. Complexity: A Gordian Knot?, in: Lovorka Galetic (Ed.): 2nd International Conference "An Enterprise Odyssey: Building Competitive Advantage" – Proceedings, Zagreb / Croatia, June 17-19, 2004, pp. 890-903.

Blecker, Thorsten / Abdelkafi, Nizar / Kaluza, Bernd / Kreutler, Gerold (2004e): A Framework for Understanding the Interdependencies between Mass Customization and Complexity, Proceedings of the 2nd International Conference on Business Economics, Management and Marketing, Athens / Greece, June 24-27, 2004.

Blecker, Thorsten / Abdelkafi, Nizar / Kreutler, Gerold (2004f): A Multi-Agent based Configuration Process for Mass Customization, in: Department of Manufacturing Engineering and Management, Technical University of Denmark (Ed.): Conference Proceedings, International Conference on Economic, Technical and Organisational aspects of Product Configuration Systems, Technical University of Denmark, Lyngby / Denmark, June 28-29, 2004, pp. 27-33.

Blecker, Thorsten / Abdelkafi, Nizar / Kreutler, Gerold / Kaluza, Bernd (2004g): Auction based Variety Formation and Steering for Mass Customization, EM - Electronic Markets, Vol. 14. No. 3 (forthcoming).

Blecker, Thorsten / Dullnig, Herwig / Malle, Franz (2003): Kundenkohaerente und kundeninhaerente Produktkonfiguration in der Mass Customization, Industrie Management, Vol. 19, No. 1, pp. 21-24.

Blecker, Thorsten / Graf, Guenter (2003): Multi Agent Systems in Internet based Production Environments – An enabling Infrastructure for Mass Customization, in: Frank T. Piller / Ralf Reichwald / Mitchell M. Tseng (Eds.): Competitive Advantage Through Customer Interaction: Leading Mass Customization and Personalization from the Emerging State to a Mainstream Business Model, Proceedings of the 2nd Interdisciplinary World Congress on Mass Customization and Personalization - MCPC'03, Munich, October 6-8, 2003, pp. 1-21.

Bliss, Christoph (2000): Management von Komplexität - Ein integrierter, systemtheoretischer Ansatz zur Komplexitätsreduktion, Wiesbaden: Gabler 2000.

Bock, Stefan / Rosenberg, Otto (2003): An Approach To Strategic Planning Of Product-Variety, in: Proceedings of the European Applied Business Research Conference 2003 - EABR'03, June 9-13, 2003, Venice/Italy, Littleton: Western Academic Press 2003, pp. 1-5.

Bond, Alan H. / Gasser, Les (1988): An Analysis of Problems and Research in DAI, in: Alan H. Bond / Les Gasser (Eds.): Readings in Distributed Artificial Intelligence. San Mateo / California: Morgan Kaufman Publishers 1988, pp. 3-35.

Bourke, Richard (2000): Product Configurators: Key Enabler for Mass Customization – An overview, URL: http://www.pdmic.com/articles/midrange/Aug2000.html (Retrieval: July 15, 2004).

Bramham, Johanna / MacCarthy, Bart L. (2003): Matching Configurator Attributes to Business Strategy, Proceedings of the MCPC 2003, 2nd Interdisciplinary World Congress on Mass Customization and Personalization, Munich, October 6-8, 2003.

Bratman, Michael E. / Israel, David J. / Pollack, Martha E. (1988): Plans and resource-bounded practical reasoning, Computational Intelligence, 1988, No. 4, pp. 349-355.

Braun, Stephan (1999): Die Prozesskostenrechnung: Ein fortschrittliches Kostenrechnungs-system?, Sternenfels / Berlin: Verlag Wissenschaft & Praxis 1999.

Broekhuizen, Thijs L. J. / Alsem, Karel, J. (2002): Success Factors for Mass Customization: A Conceptual Model, Journal of Market-Focused Management, Vol. 5, No. 4, pp. 309-330.

Brown, Stanley A. / Gulycz, Moosha (2002): Performance Driven CRM: How to Make Your Customer Relationship Management Vision a Reality, Ontario: John Wiley & Sons 2002.

Bruhn, Manfred (2002): E-Services – eine Einfuehrung in die theoretischen und praktischen Probleme, in: Manfred Bruhn / Bernd Stauss (Eds.): Electronic Services: Dienstleistungsmanagement Jahrbuch 2002, Wiesbaden: Gabler 2002.

Burke, Robin D. (2002): Hybrid recommender systems: Survey and experiments, User Modeling and User-Adapted Interaction, Vol. 12, No. 4, pp. 331-370.

Byrne, Mike D. / Chutima, Parames (1997): Real-time operational control of an FMS with full routing flexibility, International Journal of Production Economics, Vol. 51, No. 1-2, pp. 109-113.

Clark, Kim B. / Fujimoto, Takahiro (1990): The power of product integrity, Harvard Business Review, November-December 1990, pp. 107-118.

Collier, David A. (1981): The Measurement and Operating Benefits of Component Part Commonality, Decision Sciences, Vol. 12, No. 1, pp. 85-96.

Collier, David A. (1982): Aggregate Safety Stock Levels and Component Part Commonality, Management Science, Vol. 28, No. 11, pp. 1296-1303.

Corsten, Hans / Goessinger Ralf (1998): Produktionsplanung und –steuerung auf der Grundlage von Multiagentensystemen, in: Hans Corsten / Ralf Goessinger (Eds.): Dezentrale Produktionsplanungs- und -steuerungs-Systeme – Eine Einfuehrung in zehn Lektionen, Stuttgart et al.: Kohlhammer 1998, pp. 174-207.

Dahmus, Jeffrey B. / Gonzalez-Zugasti, Javier P. / Otto, Kevin N. (2001): Modular product architecture, Design Studies, Vol. 22, No. 5, pp. 409-424.

Da Silveira, Giovani / Borenstein, Denis / Fogliatto, Flávio S. (2001): Mass customization: Literature review and research directions, International Journal of Production Economics, Vol. 72, No. 1, pp. 1-13.

Davis, Stanley M. (1987): Future Perfect, Reading: Addison-Wesley Publishing 1987.

Desmeules, Rémi (2002): The impact of Variety on Consumer Happiness: Marketing and the Tyranny of Freedom, Academy of Marketing Science Review [Online] 2002, URL: http://www.amsreview.org/articles/desmeules12-2002.pdf (Retrieval: July 15, 2004).

Duray, Rebecca / Ward, Peter T. / Milligan, Glenn W. / Berry, William L. (2000): Approaches to mass customization: configurations and empirical validation, Journal of Operations Management, Vol. 18, No. 6, pp. 605-625.

Eberle, Rudolf (2000): Varianten-Controlling in der Automobilindustrie, Kostenrechnungspraxis, Vol. 44, No. 6, pp. 343-350.

Eirinaki, Magdalini / Vazirgiannis, Michalis (2003): Web Mining for Web Personalization, ACM Transactions on Internet Technology, Vol. 3, No. 1, pp. 1-27.

Ekstroem, Karin / Karlsson, MariAnne (2001): Customer Oriented Product Development? An exploratory study of four Swedish SME's, FE-rapport 2001-380, Goeteborg University 2001, URL: http://www.handels.gu.se/epc/archive/00001404/01/gunwba380.pdf (Retrieval: July 15, 2004).

Ericsson, Anna / Erixon, Gunnar (1999): Controlling Design Variants: Modular Product Platforms, Dearborn / Michigan: Society of Manufacturing Engineers 1999.

Felfernig, Alexander / Friedrich, Gerhard / Jannach, Dietmar / Zanker, Markus (2002): Web-based configuration of virtual private networks with multiple suppliers, Proceedings of the Seventh Intl. Conference on Artificial Intelligence in Design (AID'02), Cambridge, UK 2002.

Fischer, Klaus / Russ, Christian / Vierke, Gero (1998): Decision Theory and Coordination in Multiagent Systems, Research Report RR-98-02, Deutsches Forschungszentrum fuer Kuenstliche Intelligenz GmbH, October 1998.

Fisher, Marshall L. (1997): What Is the Right Supply Chain for Your Product?, Harvard Business Review, March-April 1997, pp. 105-116.

Fisher, Marshall L. / Ittner, Christopher D. (1999): The Impact of Product Variety on Automobile Assembly Operations: Empirical Evidence and Simulation Analysis, Management Science, Vol. 45, No. 6, pp. 771-786.

Forza, Cipriano / Salvador, Fabrizio (2002): Managing for variety in the order acquisition and fulfilment process: The contribution of product configuration systems, International Journal of Production Economics, Vol. 76, No. 1, pp. 87-98.

Franke, Nikolaus / Piller, Frank (2003): Key Research Issues in User Interaction with Configuration Toolkits, International Journal of Technology Management (IJTM), Vol. 26 (2003), No. 5/6, pp. 578-599.

Freuder, Eugene C. (1998): The role of configuration knowledge in the business process, IEEE intelligent systems, Vol. 13, No. 4, pp. 29-31.

Frizelle, Gerry / Efstathiou, Janet (2002): Seminar Notes On "Measuring Complex Systems", URL: http://www.psych.lse.ac.uk/complexity/PDFiles/Seminars/GerjanApril02 lastversion.pdf (Retrieval: July 15, 2004).

Frizelle, Gerry / Woodcock, Eric (1995): Measuring complexity as an aid to developing operational strategy, International Journal of operations and Production Management, Vol. 15, No. 5, pp. 26-39.

Garud, Raghu / Kumaraswamy, Arun (2003): Technological and Organizational Designs for Realizing Economies of Substitution, in: Raghu Garud / Arun Kumaraswamy / Richard N. Langlois (Eds.): Managing in the Modular Age – Architectures, Networks, and Organizations, Malden et al.: Blackwell Publishing 2003, pp. 45-77.

Goessinger, Ralf (2000): Opportunistische Koordinierung bei Werkstattfertigung, Deutscher Universitaetsverlag 2000.

Goldman, Steven / Nagel, Roger / Preiss, Kenneth (1995): Agile Competitors and Virtual Organizations – Strategies for Enriching the Customer, New York et al.: Van Nostrand Reinhold 1995.

Gruber, Thomas R. (1993): Toward Principles for the Design of Ontologies Used for Knowledge Sharing. Technical report KSL 93-04, Knowledge Systems Laboratory, Stanford University, URL: http://www-ksl.stanford.edu/knowledge-sharing/papers/onto-design.ps (Retrieval: July 15, 2004).

Guenter, Andreas / Kuehn, Christian (1999): Knowledge-Based Configuration – Survey and Future Directions – Technical report, in URL: http://www.hitec-hh.de/ueberuns/home/ aguenter/literatur/xps-99.pdf (Retrieval: July 15, 2004).

Guilabert, Margarita / Donthu, Naveen (2003): Mass Customization and Consumer Behavior: The Development of a Scale to Measure Customer Customization Sensitivity, Proceedings of the MCPC 2003, 2nd Interdisciplinary World Congress on Mass Customization and Personalization, Munich, October 6-8, 2003.

Hart, Christopher W.L. (1995): Mass customization: conceptual underpinnings, opportunities and limits, International Journal of Service Industry Management, Vol. 6, No. 2, pp. 36-45.

Hartley, John R. (1992): Concurrent Engineering – Shortening Lead times, Raising Quality, and Lowering Costs, Portland: Productivity Press 1992.

Hasselbring, Wilhelm (2000): Information System Integration, Communications of the ACM, Vol. 43, No. 6, pp. 33-38.

Hildebrand, Rudolf / Mertens, Peter (1992): PPS-Controlling mit Kennzahlen und Checklisten, Berlin – Heidelberg: Springer 1992.

Hillier, Mark S. (2000): Component commonality in multi-period, assemble-to-order systems, IIE Transactions, Vol. 32, No. 8, pp. 755-766.

Hoege, Robert (1995): Organisatorische Segmentierung – Ein Instrument zur Komplexitaetshandhabung, Wiesbaden: Gabler Verlag 1995.

Hoogeweegen, Martijn / Hagdorn-van der Meijden, Lorike (2003): Strategizing for Mass Customization by Playing the Business Networking Game, Proceedings of the MCPC 2003, 2nd Interdisciplinary World Congress on Mass Customization and Personalization, Munich, October 6-8, 2003.

Huffman, Cynthia / Kahn, Barbara E. (1998): Variety for Sale: Mass Customization or Mass Confusion, Journal of retailing, Vol. 74, No. 4, pp. 491-513.

Ishii, Kosuke (1998): Modularity: A Key Concept in Product Life-cycle Engineering, Design Division, Department of Mechanical Engineering, University of Stanford 1998, URL: http://www-mml.stanford.edu/Research/Papers/1998/1998.LEbook.ishii/1998.LEbook. ishii.pdf (Retrieval: July 15, 2004).

Ives, Blake / Learmonth, Gerard P. (1984): The information system as a competitive weapon, Communications of the ACM, Vol. 27 (1984), pp. 1193-1201.

Iyenger, Sheena S. / Lepper, Mark R. (2000): When Choice is Demotivating: Can One Desire Too Much of a Good Thing?, URL: http://www.columbia.edu/~ss957/ publications.html (Retrieval: July 15, 2004).

Jannach, Dietmar (2004): Advisor Suite – A knowledge-based sales advisory system, Proceedings of the 16th European Conference on Artificial Intelligence - 3rd Prestigious Applications Intelligent Systems Conference, Valencia, Spain 2004.

Jannach, Dietmar / Kreutler, Gerold (2004a): Building on-line sales assistance systems with ADVISOR SUITE, Proceedings of 16th International Conference on Software Engineering and Knowledge Engineering (SEKE'04), Banff, CAN 2004.

Jannach, Dietmar / Kreutler, Gerold (2004b): ADVISOR SUITE: A Tool for Rapid Development of Maintainable Online Sales Advisory Systems, Proceedings of ICWE 2004, International Conference on Web Engineering, in: Lecture Notes in Computer Science, No. 3140, Springer 2004, pp. 266-270.

Jennings, Nikolas R. (2000): On Agent-based Software Engineering, Artificial Intelligence, Vol. 117, No. 2, pp. 277-296.

Jennings, Nikolas R. / Wooldridge Michael (1998): Applications of Intelligent Agents, in: Nikolas R. Jennings / Michael Wooldridge (Eds.): Agent Technology: Foundations, Applications, and Markets, Berlin: Springer Verlag 1998, pp. 3-28.

Jiao, Jianxin / Tseng, Mitchell M. (1999): A Pragmatic Approach to Product Costing Based on Standard Time Estimation, International Journal of Operations & Production Management, Vol. 19, No. 7, pp. 738-754.

Jiao, Jianxin / Tseng, Mitchell M. (2000): Understanding Product Family for Mass Customization by Developing Commonality Indices, Journal of Engineering Design, Vol. 11, No. 3, pp. 225-243.

Juengst, Werner E. / Heinrich, Michael (1998): Using Resource Balancing to Configure Modular Systems, IEEE intelligent systems, Vol. 13, No. 4, pp. 50-58.

Jugel, Albert (2003): Markterfolg durch differenzierte kundenwunschspezifische Leistungen, in: Horst Wildemann (Ed.): Fuehrungsverantwortung – Bewaehrte oder innovative Managementmethoden?, Tagungsband Muenchner Management Kolloquium, Muenchen: TCW Transfer-Centrum, pp. 411-438.

Kahn, Barbara (1998): Variety: From the Consumer's perspective, in: Teck-Hua Ho / Christopher S. Tang (Eds.): Product Variety Management – Research Advances, Boston et al.: Kluwer Academic Publishers 1998, pp. 19-37.

Kaluza, Bernd (1989): Erzeugniswechsel als unternehmenspolitische Aufgabe – Integrative Loesungen aus betriebswirtschaftlicher und ingenieurwissenschaftlicher Sicht, Berlin: Erich Schmidt 1989.

Kaluza, Bernd (1995): Flexibilität der Industrieunternehmen, Working Paper Nr. 208, Department of Economics, University of Duisburg / Germany, March 1995.

Kaplan, Robert S. / Norton, David P. (1997): The Balanced Scorecard – Strategien erfolgreich umsetzen, Stuttgart: Schaeffer-Poeschel Verlag 1997.

Kauffman, Stuart A. (1993): The Origins of Order: Self-Organization and Selection in Evolution, New York: Oxford University Press 1993.

Kirn, Stefan (2002): Kooperierende intelligente Softwareagenten, Wirtschaftsinformatik, Vol. 44, No. 1, pp. 53-63.

Klemperer, Paul (1999): Auction Theory: A Guide to the Literature, Journal of Economic Surveys, Vol. 13, No. 3, pp. 227-286.

Klusch, Matthias (1999): Intelligent Information Agents, Berlin: Springer 1999.

Kobsa, Alfred / Koenemann, Juergen / Pohl, Wolfgang (2001): Personalized Hypermedia Presentation Techniques for Improving Online Customer Relationships, The Knowledge Engineering Review, Vol. 16, No. 2, pp. 11-155.

Kotha, Suresh (1995): Mass Customization: Implementing the Emerging Paradigm for Competitive Advantage, Strategic Management Journal, Vol. 16, Special Issue, pp. 21-42.

Kotha, Suresh (1996): From mass production to mass customization: The case of the National Industrial Bicycle Company of Japan, European Management Journal, Vol. 14, No. 5, pp. 442-450.

Krallmann, Hermann / Albayrak, Sahin (2001): Intelligente Agenten zur Steuerung von dezentralen Fertigungsstrukturen, in: Klaus Bellmann (Ed.): Kooperations- und Netzwerkmanagement: Festgabe fuer Gert v. Kortzfleisch zum 80. Geburtstag, Berlin: Duncker und Humblot 2001, pp. 111-128.

Krasner, Glenn E. / Pope, Stephen T. (1988): A Description of the Model-View-Controller User Interface Paradigm in the Smalltalk-80 System, ParcPlace Systems Inc., Mountain View 1988.

Kuepper, Hans-Ulrich (2001): Controlling: Konzeption, Aufgaben und Instrumente, Stuttgart: Schaeffer-Poeschel Verlag 2001.

Langlois, Richard N. / Robertson, Paul L. (2003): Networks and Innovation in a Modular System: Lessons From the Microcomputer and Stereo Component industries, in: Raghu Garud / Arun Kumaraswamy / Richard N. Langlois (Eds.): Managing in the Modular Age – Architectures, Networks, and Organizations, Malden et al.: Blackwell Publishing 2003, pp. 78-113.

Leckner, Thomas / Stegmann, Rosmary / Schlichter, Johann (2004): Reducing Complexity for Customers by means of a Model-Based Configurator and Personalized Recommendations, in: Department of Manufacturing Engineering and Management, Technical University of Denmark (Ed.): Conference Proceedings, International Conference on Economic, Technical and Organisational aspects of Product Configuration Systems, Technical University of Denmark, Lyngby / Denmark, June 28-29, 2004, pp. 199-208.

Lingnau, Volker (1994): Variantenmanagement: Produktionsplanung im Rahmen einer Produktdifferenzierungsstrategie, Berlin: Erich Schmidt Verlag 1994.

MacCarthy, Bart L. / Brabazon, Philip G. / Bramham, Johanna (2002): Key Value Attributes in Mass Customization, in: Claus Rautenstrauch / Ralph Seelmann-Eggebert / Klaus Turowski (Eds.): Moving into Mass Customization: Information Systems and Management Principles, Berlin – Heidelberg: Springer 2002, pp. 71-87.

MacCarthy, Bart L. / Brabazon, Philip G. / Bramham, Johanna (2003): Fundamental modes of operation for mass customization, International Journal of Production Economics, Vol. 85, No. 3, pp. 289-304.

Mailharro, Daniel (1998): A classification and constraint-based framework for configuration, Artificial Intelligence for Engineering Design, Analysis and Manufacturing, Vol. 12, No. 4, pp. 383-397.

Maroni, Dirk (2001): Produktionsplanung und -steuerung bei Variantenfertigung, Frankfurt am Main et al.: Peter Lang 2001.

Martin, Mark V. / Ishii, Kosuke (1996): Design For Variety: A Methodology For Understanding the Costs of Product Proliferation, Proceedings of The 1996 ASME Design Engineering Technical Conferences and Computers in Engineering Conference, California, August 18-22, 1996, URL: http://mml.stanford.edu/Research/Papers/1996/1996.ASME.DTM.Martin/1996.ASME.DTM.Martin.pdf (Retrieval: July 15, 2004).

Martin, Mark V. / Ishii, Kosuke (1997): Design For Variety: Development of Complexity Indices and Design Charts, Proceedings of DETC'97, 1997 ASME Design Engineering Technical Conferences, Sacramento, September 14-17, 1997, URL: http://mml.stanford.edu/Research/Papers/1997/1997.ASME.DFM.Martin/1997.ASME.DFM.Martin.pdf (Retrieval: July 15, 2004).

Maskell, Brian (1991): Performance Measurement for World Class Manufacturing: A Model for American Companies, Cambridge / Massachusetts: Productivity Press 1991.

Matzler, Kurt / Hinterhuber, Hans H. (1998): How to make product development projects more successful by integrating Kano's model of customer satisfaction into quality function deployment, Technovation, Vol. 18, No. 1, pp. 25-38.

Mchunu, Claudia / de Alwis, Aruna / Efstathiou, Janet (2003): Decision support framework for establishing a « best fit » mass customization strategy, Working Paper, University of Oxford, 2003, URL: http://www.robots.ox.ac.uk/~manufsys/mcu/papers/ijppc/bestfit001.pdf (Retrieval: July 15, 2004).

Meissner, Harald (2002): Elektronische Kundendialoge als Element elektronischer Dienstleistungen – Vorschlag einer Systematisierung und Ableitung von Konsequenzen fuer Automatisierung und Selbstbedienung, in: Manfred Bruhn / Bernd Strauss (Eds.): Electronic Services: Dienstleistungsmanagement Jahrbuch 2002, Wiesbaden: Gabler 2002.

Meyer, Marc / Lehnerd, Alvin (1997): The Power of Product Platforms: Building Value and Cost Leadership, New York: The Free Press 1997.

Mikkola, Juliana H. / Gassmann, Oliver (2003): Managing Modularity of Product Architectures: Toward an Integrated Theory, IEEE Transactions on Engineering Management, Vol. 50, No. 2, pp. 204-218.

Miller, George A. (1956): The Magical Number Seven, Plus or Minus Two: Some Limits on Our Capacity for Processing Information, The Psychological Review, Vol. 63, pp. 81-97, URL: http://www.well.com/user/smalin/miller.html (Retrieval: July 15, 2004).

Moulin, Bentahar / Chaib-draa, Brahim (1996): An Overview of Distributed Artifical Intelligence, in: Gregory M. P. O'Hare / Nikolas R. Jennings (Eds.): Foundations of Distributed Artificial Intelligence, New York: John Wiley & Sons 1996, pp. 3-56.

Mueller, Volkmar (2001): Konzeptionelle Gestaltung des operativen Produktionscontrolling unter Beruecksichtigung von differenzierten Organisationsformen der Teilefertigung, Aachen: Shaker Verlag 2001.

Muther, Andreas (2000): Electronic Customer Care: Die Anbieter-Kunden-Beziehung im Informationszeitalter, 2nd Edition, Berlin: Springer 2000.

Nagamachi, Mitsuo (1995): Kansei Engineering: A new ergonomic consumer-oriented technology for product development, International Journal of Industrial Ergonomics, Vol. 15, No. 1, pp. 3-11.

Nagamachi, Mitsuo (2002): Kansei engineering as a powerful consumer-oriented technology for product development, Applied Ergonomics, Vol. 33, No. 3, pp. 289-294.

Naylor, J. Ben / Naim, Mohamed M. / Berry, Danny (1999): Leagility: integrating the lean and agile manufacturing paradigms in the total supply chain, International journal of Production Economics, Vol. 62, No.1-2, pp. 107-118.

Neely, Andy (1999): The performance measurement revolution: why now and what next?, International Journal of Operations & Production Management, Vol. 19, No. 2, pp. 205-228.

Neely, Andy / Gregory, Mike / Platts, Ken (1995): Performance measurement system design – A literature review and research agenda, International Journal of Operations & Production Management, Vol. 15, No. 4, pp. 80-116.

Nilles, Volker (2002): Effiziente Gestaltung von Produktordnungssystemen – Eine theoretische und empirische Untersuchung, Ph.D. Dissertation, Technische Universitaet Muenchen, Muenchen 2002.

Nwana, Hyacinth S. / Ndumu, Divine T. (1998): A Brief Introduction to Software Agent Technology, in: Nikolas R. Jennings / Michael Wooldridge (Eds.): Agent Technology: Foundations, Applications, and Markets, Berlin: Springer Verlag 1998, pp. 29-48.

Oleson, John D. (1998): Pathways to Agility: Mass Customization in Action, New York et al.: John Wiley & Sons 1998.

Paulson, Patrick / Tzanavari, Aimilia (2002): Combining Collaborative and Content-Based filtering Using Conceptual Graphs. Technical Report TR-2002-5, Department of Computer Science, University of Cyprus, December 2002, URL: http://citeseer.nj.nec.com/ 579370.html (Retrieval: July 15, 2004).

Perridon, Louis / Steiner, Manfred (1999): Finanzwirtschaft der Unternehmung, 10th Edition, Muenchen: Vahlen 1999.

Piller, Frank T. (1998): Kundenindividuelle Massenproduktion – Die Wettbewerbsstrategie der Zukunft, Muenchen, Wien: Hanser 1998.

Piller, Frank T. (2000): Mass Customization – Ein wettbewerbsstrategisches Konzept im Informationszeitalter, Wiesbaden: Gabler Verlag 2000.

Piller, Frank T. (2002): Logistische Kennzahlen und Einflussgroessen zur Performance-Bewertung der Mass-Customization-Systeme von Selve und Adidas, Working Paper, Department of General and Industrial Management, TUM Business School, Muenchen 2002.

Piller, Frank T. / Ihl, Christoph (2002): Mythos Mass Customization: Buzzword oder praxisrelevante Wettbewerbsstrategie? – Warum viele Unternehmen trotz der Nutzenpotentiale Kundenindividueller Massenproduktion an der Umsetzung scheitern, Working paper, Department of Information, Organization and Management, Technische Universitaet Muenchen, URL: http://www.mass-customization.de/mythos.pdf (Retrieval: July 15, 2004).

Piller, Frank T. / Koch, Michael / Moeslein, Kathrin / Schubert, Petra (2003): Managing High Variety: How to Overcome the Mass Confusion Phenomenon, Proceedings of the EURAM 2003 Conference, Milan, April 2003, URL: http://dwi.fhbb.ch/eb/publications. nsf/0/be03b7c17ed2d9b0c1256ccb00346e3a/$FILE/euram-mass-confusion.pdf (Retrieval: July 15, 2004).

Piller, Frank T. / Reichwald, Ralf (2002): Mass Customization, in: Zheng Li / Frank Possel-Doelken (Eds.): Strategic Production Networks, New York et al.: Springer 2002, pp. 389-421, URL: http://www.mass-customization.de/download/spn2000.pdf (Retrieval: July 15, 2004).

Piller, Frank T. / Tseng, Mitchell M. (2003): New Directions for Mass Customization. Setting an agenda for future research and practice in mass customization, personalization, and customer integration, in: Frank T. Piller / Mitchell M. Tseng (Eds.): The Customer Centric Enterprise: Advances in Mass Customization and Personalization, New York - Berlin, pp. 513 - 529.

Piller, Frank T. / Waringer, Daniela (1999): Modularisierung in der Automobilindustrie – neue Formen und Prinzipien, Aachen: Shaker Verlag 1999.

Pine II, B. Joseph (1993): Mass Customization: The New Frontier in Business Competition, Boston, Massachusetts: Harvard Business School Press 1993.

Pine II, B. Joseph (2003): An introduction by B. Joseph Pine II, Proceedings of the MCPC 2003, 2nd Interdisciplinary World Congress on Mass Customization and Personalization, Munich, October 6-8, 2003.

Pine II, B. Joseph / Gilmore, James H. (1999): The Experience Economy, Boston, Massachusetts: Harvard Business School Press 1993.

Porter, Michael E. (1998): Competitive Advantage – Creating and Sustaining Superior Performance, 2nd Edition, New York et al.: The Free Press 1998.

Rao, Anand S. / Georgeff, Michael P. (1991): Modeling rational agents within a BDI-architecture, in: Richard Fikes / Eric Sandewall (Eds.): Proceedings of Knowledge Representation and Reasoning (KR&R-91), Morgan Kaufmann 1991, pp. 473-484.

Rautenstrauch, Claus (1997): An alternative concept to MRPII for mass customization, Proceedings of the international Conference of Manufacture Automation 1997, The University of Hong Kong, pp. 401-406.

Rautenstrauch, Claus / Taggermann, Holger / Turowski, Klaus (2002): Manufacturing Planning and Control Content Management in Virtual Enterprises Pursuing Mass Customization, in: Claus Rautenstrauch / Ralph Seelmann-Eggebert / Klaus Turowski (Eds.): Moving into Mass Customization: Information Systems and Management Principles, Berlin Heidelberg: Springer-Verlag 2002, pp. 103-118.

Reichmann, Thomas (2001): Controlling mit Kennzahlen und Managementberichten: Grundlagen einer systemgestuetzten Controlling-Konzeption, Muenchen: Vahlen 2001.

Reichwald, Ralf (1993): Kommunikation, in: Michael Bitz / Klaus Dellmann / Michael Domsch / Henning Egner (Eds.): Vahlens Kompendium der Betriebswirtschaftslehre, Muenchen: Vahlen 1993, pp. 447-494.

Reiss, Michael (1992): Optimieren der Unternehmenskomplexitaet, io Management Zeitschrift, Vol. 61, No.7/8, pp. 40-43.

Reiss Michael / Beck, Thilo C. (1994): Fertigung jenseits des Kosten-Flexibilitaets-Dilemmas. Mass Customization als Strategiekonzept fuer Massenfertiger und Einzelfertiger, VDI-Z, Vo. 136, No. 11/12, pp. 28-30.

Resnick, Paul / Varian, Hal R. (1997): Recommender Systems, Communications of the ACM, Vol. 40, No. 3, pp. 56-58.

Riemer, Kai / Totz, Carsten (2001): The many faces of personalization – An integrative economic overview of mass customization and personalization, Proceedings of the MCPC 2001, 1st Interdisciplinary World Congress on Mass Customization and Personalization, Hong Kong, October 1-2, 2001.

Riquelme, Hernan (2001): Do consumers know what they want?, Journal of Consumer Marketing, Vol. 18, No. 5, pp. 437-448.

Robben, Matthias (2001): Service Sells - Kundenberatung online, URL: http://www.ecin.de/ strategie/kundenberatung (Retrieval: July 15, 2004).

Rogoll, Timm / Piller, Frank (2002): Konfigurationssysteme fuer Mass Customization und Variantenproduktion, Muenchen: ThinkConsult 2002.

Rosenberg, Otto (1996): Variantenfertigung, in: Werner Kern / Hans-Horst Schroeder / Juergen Weber (Eds.): Handwoerterbuch der Produktionswirtschaft, 2nd Edition, Stuttgart: Schaeffer-Poeschel 1996, pp. 2119-2129.

Rosenberg, Otto (2002): Kostensenkung durch Komplexitaetsmanagement, in: Klaus-Peter Franz / Peter Kajueter (Eds.): Kostenmanagement: Wertsteigerung durch systematische Kostensenkung, Stuttgart: Schaeffer Poeschel Verlag 2002, pp. 225-245.

Rosenschein, Jeffrey S. (1985): Rational Interaction: Cooperation Among intelligent Agents. PhD thesis, Computer Science Department, Stanford University, Stanford 1985.

Russel, Stuart / Norvig, Peter (1995): Artificial Intelligence: A Modern Approach, New Jersey: Prentice-Hall, Inc. 1995.

Sabin, Daniel / Weigel, Rainer (1998): Product Configuration Frameworks – A Survey, IEEE intelligent systems, Vol. 13, No. 4, pp. 42-49.

Saeed, Barry / Young, David (1998): Managing the Hidden Costs of Complexity, Boston Consulting Group, Whitepaper, URL: http://www.healy-hudson.com/_ADD_ON/_download/Managing_hidden_costs.pdf (Retrieval: July 15, 2004).

Salvador, Fabrizio / Forza, Cipriano / Rungtusanatham, Manus (2002a): Modularity, product variety, production volume, and component sourcing: theorizing beyond generic prescriptions, Journal of operations management, Vol. 20, No. 5, pp. 549-575.

Salvador, Fabrizio / Forza, Cipriano / Rungtusanatham, Manus (2002b): How to mass customize: Product architectures, sourcing configurations, Business Horizons, Vol. 45, No. 4, pp. 61-69.

Sanchez, Ron / Collins, Robert P. (2001): Competing-and Learning-in Modular Markets, Long Range Planning, Vol. 34, No. 6, pp. 645-667.

Sandholm, Tuomas W. (1999): Automated negotiation, Communications of the ACM, Vol. 42, No. 3, pp. 84-85.

Schaefer, Hans F. (1993): Anwendungsgrenzen der Prozesskostenrechnung bei komplexen Produktionsprozessen, Kostenrechnungspraxis, Vol. 37, No. 5, pp. 310-313.

Schaefer, Ralph (2001): Rules for Using Multi-Attribute Utility Theory for Estimating a User's Interests, Proceedings des 9. GI-Workshops: ABIS-Adaptivitaet und Benutzermodellierung in interaktiven Softwaresystemen, October 8-10, 2001.

Schafer, Ben J. / Konstan, Joseph A. / Riedl, John (1999): Recommender Systems in E-Commerce, ACM Conference on Electronic Commerce, 1999, pp. 158-166, URL: http://www.grouplens.org/papers/pdf/ec-99.pdf (Retrieval: July 15, 2004).

Scheer, Christian / Hansen, Torben / Loos, Peter (2003): Erweiterung von Produktkonfiguratoren im Electronic Commerce um eine Beratungskomponente. ISYM-Working Paper, Nr. 11, Johannes Gutenberg-Universitaet Mainz, Lehrstuhl Wirtschaftsinformatik und Betriebswirtschaftslehre, Mainz, URL: http://chris.scheer.bei.t-online.de/docs/scheer03_konfigurator_beratung.pdf (Retrieval: July 15, 2004).

Schenk, Michael / Seelmann-Eggebert, Ralph (2003): Mass Customization across the Value Chain, Proceedings of the MCPC 2003, 2nd Interdisciplinary World Congress on Mass Customization and Personalization, Munich, October 6-8, 2003.

Schuetz, Wilken / Meyer, Markus (2001): Definition einer Parameter-Hierarchie zur Adaptierung der Benutzer-Interaktion in E-Commerce-Systemen, Proceedings des 9. GI-Workshops: ABIS-Adaptivitaet und Benutzermodellierung in interaktiven Softwaresystemen, October 8-10, 2001.

Schuh, Guenther / Schwenk, Urs (2001): Produktkomplexitaet managen, Muenchen, Wien: Carl Hanser Verlag 2001.

Schwartz, Barry (2000): Self-Determination: The Tyranny of Freedom, URL: http://www.swarthmore.edu/SocSci/bschwar1/self-determination.pdf (Retrieval: July 15, 2004).

Seidenschwarz, Werner (2001): Target Costing. Marktorientiertes Zielkostenmanagement, Muenchen: Vahlen 2001.

Shehory, Onn / Kraus, Sarit (1995): Coalition formation among autonomous agents: Strategies and complexity, in: Christiano Castelfranchi / Jean Pierre Muller (Eds.): From Reaction to Cognition, Heidelberg: Springer Verlag 1995, pp. 57-72.

Shehory, Onn / Sycara, Katia P. / Jha, Somesh (1997): Multi-agent Coordination through Coalition Formation, in: Intelligent Agents IV: Agent Theories, Architectures and Languages, Lecture Notes in Artificial Intelligence, No. 1365, Springer 1997, pp. 143-154.

Siddique, Zahed (2000): Common Platform Development: Designing for Product Variety, Ph.D. Dissertation, Georgia Institute of Technology, URL: http://srl.marc.gatech.edu/people/zahed/PhD/PhDIndex.htm (Retrieval: July 15, 2004)

Siddique, Zahed / Rosen, David W. (2001): Identifying Common Platform Architecture for a Set of Similar Products, Proceedings of the MCPC 2001, 1st Interdisciplinary World Congress on Mass Customization and Personalization, Hong Kong, October 1-2, 2001.

Slack, Nigel (1983): Flexibility as a Manufacturing Objective, International Journal of Operations and Production Management, Vol. 3, No. 3, pp. 4-13.

Smith, Reid G. / Davis, Randall (1980): Frameworks for cooperation in distributed problem solving, IEEE Transactions on Systems, Man and Cybernetics, Vol. 11, No. 1, pp. 61-70.

Soininen, Timo / Tiihonen, Juha / Männistö, Tomi / Sulonen, Reijo (1998): Towards a general ontology of configuration, Artificial Intelligence for Engineering Design, Analysis and Manufacturing, Vol. 12, No. 4, pp. 357-372.

Sterne, Jim (2002): Web Metrics: Proven Methods for Measuring Web Site Success, New York: John Wiley & Sons 2002.

Stone, Robert B. / Wood, Kristin L. / Crawford, Richard H. (2000): A heuristic method for identifying modules for product architectures, Design Studies, Vol. 21, No. 1, pp. 5-31.

Stroebel, Michael (2000): On Auctions as the Negotiation Paradigm of Electronic Markets, Electronic Markets, Vol. 10, No. 1, pp. 39-44.

Svensson, Carsten / Jensen, Thomas (2001): The customer at the final frontier of mass customisation, Proceedings of the MCPC 2001, 1st Interdisciplinary World Congress on Mass Customization and Personalization, Hong Kong, October 1-2, 2001.

Tersine, Richard J. / Wacker, John G. (2000): Customer-Aligned Inventory Strategies: Agility Maxims, International Journal of Agile Management Systems, Vol. 2, No. 2, pp. 114-120.

Tiihonen, Juha / Soininen, Timo (1997): Product Configurators – Information System Support for Configurable Products, in: Increasing Sales Productivity through the Use of Information Technology during the Sales Visit A survey of the European Market, Hewson Consulting Group 1997.

Toffler, Alvin (1980): The Third Wave, New York: William Morrow & Co., Inc. 1980.

Treleven, Mark / Wacker, John G. (1987): The Sources, Measurements, and Managerial Implications of Process Commonality, Journal of Operations Management, Vol. 7, No. 1/2, pp. 11-25.

Tsang, Edward (1993): Foundations of Constraint Satisfaction, London: Academic Press 1993.

Tseng, Mitchell M. / Jiao, Jianxin (2001): Mass customization, in: Gavriel Salvendy (Ed.): Industrial Engineering Handbook: Technology and Operations Management, 3rd Edition, New York: John Wiley and Sons 2001, pp. 684-709, URL: http://ami.ust.hk/ami/public/publication/Book%20Chapters/Chap25MassCustomization.pdf (Retrieval: July 15, 2004).

Turowski, Klaus (2002): Agent-based e-commerce in case of mass customization, International Journal of Production Economics, Vol. 75. No. 1-2, pp. 69-81

Ulrich, Hans / Probst, Gilbert J. B. (1995): Anleitung zum ganzheitlichen Denken und Handeln, 4th Edition, Bern Stuttgart: Paul Haupt Verlag 1995.

Ulrich, Karl (1995): The role of product architecture in the manufacturing firm, Research Policy, Vol. 24, No. 3, pp. 419-440.

Ulrich, Karl / Eppinger, Steven (2000): Product design and development, 2nd Edition, Boston: McGraw-Hill 2000.

Ulrich, Karl / Tung, Karen (1991): Fundamentals of product modularity, Proceedings of the 1991 ASME Winter Annual Meeting Symposium on Issues in Design / Manufacturing Integration, New York: American Society of Mechanical Engineers.

van Hoek, Remko I. (2001): The rediscovery of postponement a literature review and directions for research, Journal of Operations Management, Vol. 19, No. 2, pp. 161-184.

Virens, M. / Hofstede, F. T. (2000): Linking Attributes, Benefits and Consumer Values. Marketing Research, URL: http://www.intelliquest.com/resources/published/MVriens1.pdf (Retrieval: July 15, 2004).

von Hippel, Eric (2001): User Toolkits for Innovation, Working paper, MIT Sloan School of Management, URL: http://ebusiness.mit.edu/research/papers/134%20vonhippel,%20Toolkits.pdf (Retrieval: July 15, 2004).

Walczuch, Rita M. / Hofmaier, Katja (1999): Measuring Customer Satisfaction on the Internet, Proceedings of RSEEM 1999, URL: http://www-edocs.unimaas.nl/files/rm00051.pdf (Retrieval: July 15, 2004).

Waller, Matthew A. / Dabholkar, Pratibha A. / Gentry Julie J. (2000): Postponement, Product Customization, and Market-Oriented Supply Chain Management, Journal of Business Logistics, Vol. 21, No. 2, pp. 133-136.

Wigand, Rolf / Picot, Arnold / Reichwald, Ralf (1997): Information, Organization and Management, Chichester et al: John Wiley & Sons 1997.

Weber, Juergen / Kummer, Sebastian / Großklaus, Armin / Nippel, Harald / Warnke, Dorothée (1995): Methodik zur Generierung von Logistik-Kennzahlen, in: Juergen Weber (Ed.): Kennzahlen fuer die Logistik, Stuttgart: Schaeffer-Poeschel Verlag 1995, pp. 9-45.

Weber, Juergen / Schaeffer, Utz (1999): Entwicklung von Kennzahlensystemen, Working paper Nr. 62, WHU Koblenz, URL: http://www.marketing.uni-essen.de/studium/vorlesungen-uebungen/Marketing-Management/2003-05-23_Kennzahlensysteme.pdf (Retrieval: July 15, 2004).

Weiss, Gerhard (1999): Prologue, in: Gerhard Weiss (Ed.): Multiagent Systems. A Modern Approach to Distributed Artificial Intelligence, Cambridge / Massachusetts: MIT Press 1999, pp. 1-23.

Wiederhold, Gio (1992): Mediators in the architecture of future information, IEEE Computer, March 1992, pp. 38-49.

Wildemann, Horst (1995a): Das Just-In-Time Konzept: Produktion und Zulieferung auf Abruf, 4th Edition, Muenchen: TCW Transfer-Centrum 1995.

Wildemann, Horst (1995b): Produktionscontrolling: Systemorientiertes Controlling schlanker Produktionsstrukturen, 2nd Edition, Muenchen: TCW Transfer-Centrum 1995.

Wildemann, Horst (1997): Logistik Prozessmanagement, Muenchen: TCW Transfer-Centrum 1997.

Wildemann, Horst (2000): Komplexitaetsmanagement: Vertrieb, Produkte, Beschaffung, F&E, Produktion und Administration, Muenchen: TCW Transfer-Centrum 2000.

Wildemann, Horst (2003): Produktordnungssysteme: Leitfaden zur Standardisierung und Individualisierung des Produktprogramms durch intelligente Plattformstrategien, Muenchen: TCW Transfer-Centrum 2003.

Wooldridge, Michael (1999): Intelligent Agents, in: Gerhard Weiss (Ed.): Multiagent Systems. A Modern Approach to Distributed Artificial Intelligence, Cambridge / Massachusetts: MIT Press 1999, pp. 27-78.

Wooldridge, Michael (2000): Reasoning about Rational Agents, Cambridge et al.: MIT Press 2000.

Wooldridge, Michael (2002): An Introduction to MultiAgent Systems, Chichester: John Wiley & Sons 2002.

Wooldridge, Michael / Jennings, Nikolas R. (1995): Intelligent agents: theory and practice, The Knowledge Engineering Review, Vol. 10, No. 2, pp. 115-152.

Yang, Sun-mo / Nagamachi, Mitsu / Lee, Soon-yo (1999): Rule-based inference model for the Kansei Engineering System, International Journal of Industrial Ergonomics, Vol. 24, No. 5, pp. 459-471.

Zipkin, Paul (2001): The limits of mass customization, Sloan Management Review, Vol. 42, No. 3, pp. 81-87.

Zwicky, Fritz (1966): Entdecken, Erfinden, Forschen im Morphologischen Weltbild. Muenchen, Zuerich, 1966.

Index

abortion rate 204, 220
achievement potential 12, 65, 182, 203
activity based costing 175, 183, 184, 193, 241
advisory component 103, 132, 144
advisory quality 94
advisory system 79, 93, 97–98, 102–12
agent pool 129, 150
agility 40
Anderson, David 35, 37, 53, 183, 189, 191, 194, 195, 196, 205, 249
artificial intelligence 117–19
Ashby, Ross W. 47, 48, 249
assembler 16, 83
attribute 229
    basic 232
    excitement 232
    level 236, 239
    performance 232
auction 133, 135
auction agent 131, 135–36, 144
automotive industry 51, 52, 137, 236

back end system 49, 50, 56
blackboard systems 125
building oriented system 50
business strategy 2, 45, 115

capacity utilization 209, 224
CAWICOMS 85

changeover 35, 192, 207
churn rate 237, 238
Cmax.com 51
coalition 130, 131, 133, 134, 137
co-designer 54, 219
collaborative filtering 99
commonality
    part 194–200
    process 191, 210–12
competitive advantage 11–12, 25, 46, 53, 192
complaint rate 237
complexity 46
    breaker 59–60
    control 47
    driver 59–60
    dynamic 46
    external 115, 149, 151
    internal 115, 149, 151
    prevention 47
    reduction 47
    structural 46
    system 46
Computer Aided Design 84
Computer Aided Planning 215
Computer Integrated Manufacturing 215
concurrent engineering 167, 168
configuration
    case-based 83
    database 173
    knowledge base 81

logic 81, 87, 173, 214
    model-based 82
    product-oriented 151
    rule-based 81
    system 80, 90, 150, 153, 155, 214
    task 55, 80, 81, 82, 85, 86, 152
configurator *See configuration system*
content based filtering 99
contract nets 125
contribution margin accounting 116, 183,
    241
coordination process 146, 150
co-producer 3, 54
CORBA 155, 158, 159
Corsten, Hans 125
Customer Buying Cycle 94–96
customer happiness 234, 237, 238, 239
customer interests modeling 108, 110,
    112
customer needs' elicitation 27, 244
customer relationship management 3, 89,
    95, 145
customer satisfaction 67, 70, 71, 73, 95,
    168, 190, 203, 232–34, 237, 244, 256
customers' needs
    model 65, 69, 72
    objective 64–72, 97, 100, 107, 232,
        237
    subjective 64–72, 97, 100, 107, 237
customization
    adaptive 14
    collaborative 14
    cosmetic 14
    hard 16, 17
    point-of-delivery 16
    self 16
    service 16
    soft 16
    transparent 14
customization system 50
customization-standardization mix 17

Da Silveira, Giovani 18
decoupling interface 154
decoupling point 50, 193, 223
delivery time reliability 213
Dell Computer 61
dialog component 103
differentiation index 223, 248

differentiation point 205, 206, 207, 223,
    224
Duray, Rebecca 15

economies of
    scale 2, 5, 9, 12, 163, 182
    scope 5
    substitution 5, 168, 170, 171, 179
Enterprise Requirement Planning 215
entropic measurement 49

fabricator 16, 83
flexibility 191, 192, 201, 207, 210
flexible customization 17
Freuder, Eugene C. 80
front end system 49, 50
functional oriented system 50

Gilmore, James H. 13
graphical user interface 103

hidden design parameters 172
hidden information 172

information supply and need model 63–
    64
intelligent agent 120–23
interface complexity 203, 217–19, 239
internal abilities 30, 36
involver 16, 83

Java 104, 106, 111
job shop manufacturing 12, 43

Kano's model 232–33
Kansei Engineering 108, 109–11, 227–
    29, 262
key metrics 184–86, 193
key metrics system 184, 221, 226, 227,
    241
key value attributes 32, 229–31
knowledge acquisition component 103
Kotha, Suresh 25

lead time 169, 191, 203, 208–9

MacCarthy, Bart L. 19
make-to-order 58, 241

make-to-stock 9
Manufacturing Resource Planning 54
manufacturing system 49
manufacturing-related tasks 2, 42, 146,
    242, 247
markets of one 1, 46, 61
mass customization 2, 12, 40–42
    modes 19
    strategies 13, 15, 16, 18, 43
    web-based 4, 234, 244
mass production 9–10
modular architecture 164
Modular Function Deployment 174–76
modularity 163
    benefits of 168–69
    bus 165
    combinatorial 165
    limits of 169–70
    managerial implications 170–73
    product 164
    sectional 165
    slot 165
modularity matrix 178
modularization 165
modularizer 16, 83
module agent 129, 137–44, 144
    account 130, 131, 138, 145
    self-preservation 129
    task 130
module drivers 175
morphological box 4, 79, 81, 90, 113, 244
multi-agent system 123–27
multiple use 202, 217, 218, 219, 239

National Industrial Bicycle Company of
    Japan (NIBC) 25

online analytical processing 145, 158

Pareto analysis 183, 241
performance measurement 185–86
personalization 98, 102, 103, 108–11,
    257, 258
Piller, Frank T. 41
Pine, Joseph B. 2, 9, 10, 12, 13, 23, 24,
    25
platform agent 129, 136–37, 144
platform efficiency 217–20, 239
process simplification 108–11

product architecture 164
product arrangement system 49, 50, 53,
    61, 167
product constraints' agent 131, 158
product customization 2, 9, 10–12
Product Data Management 84, 215
product family 19, 37, 80, 133, 172, 176,
    177, 178, 183, 194, 196, 197, 199,
    200, 211, 241
product orientation 92
Production Planning and Control 54
prosumer 54

rationality 118, 130, 139
reaching agreement 133, 245
recommender systems 98–101
    attribute based 99
    collaborative 99
    non-personalized 99
    personalized 99
Reichwald, Ralf 70
return rate 237, 239

Sabin, Daniel 80, 81, 82, 90, 259
setup 35, 207, 208, 210, 212
solution space 12, 46, 65, 71, 74, 82–84,
    89, 93, 109, 115, 116, 127, 147, 219,
    244
standardization 9, 17, 18, 19, 21, 56, 58,
    59, 61, 97, 158, 166, 182, 200, 202,
    205
sub-process
    development 189, 194
    information 193, 213
    interaction 190, 203
    logistic 192, 212
    production 191, 205
    purchasing 190, 205
success factors 23, 247
supply chain agility 28, 36, 39, 40

target costing 132
target costing agent 131, 144
time pattern analysis 216, 217
Tseng, Mitchell M. 54, 56, 197, 198, 199,
    200, 204, 210, 211, 223, 250, 251,
    254, 257, 260

validation 109, 110, 113

validation agent 132, 144
value chain readiness 33
variety
    formation 116, 131, 144
    management 182
    perceived 203, 220, 221, 234, 240
    possible 56, 219, 220, 221
    steering 116, 131, 144, 182, 184, 187, 218, 242
vertical integration 9
visible design rules 179
visible information 172
Volkswagen 137

web metrics 108, 110, 112
web mining 108, 110, 112, 245
Weigel, Rainer 80, 81, 82, 90, 259
Wildemann, Horst 47, 52–53, 167
work-in-process 58, 188, 192, 207, 212, 222, 226

XML 85, 111, 155, 156, 158

Authors

Nizar Abdelkafi

is a PhD candidate and research follow at the Department of Production/Operations Management, Business Logistics and Environmental Management at the University of Klagenfurt, Austria within the interdisciplinary multi-year research projects "Modeling, Planning, and Assessment of Business Transformation Processes in the Area of Mass Customization" and "TECTRANS – Technology Transfer". He holds an industrial engineering diploma from the National Engineering School of Tunis, Tunisia, and a Master in Business Administration from the Technische Universität München, Germany. He has participated in several international conferences and workshops in the areas of business administration and engineering. Main research interests: competitive strategies, especially mass customization, variety and complexity management as well as the application of agent technology in practice.

email: nizar.abdelkafi@uni-klu.ac.at
Homepage: http://www.uni-klu.ac.at/plum/

Thorsten Blecker

is associate professor at the Department of Production/Operations Management, Business Logistics and Environmental Management, University of Klagenfurt, Austria. He holds a masters degree in business administration (with honors) and a PhD (summa cum laude) from the University of Duisburg, Germany. He finished his habilitation thesis in September 2004 at the University of Klagenfurt, Austria. Thorsten Blecker is guest-editor of a special issue of IEEE Transactions on Engineering Management on "Mass Customization Manufacturing Systems" (forthcoming), co-editor of the books "Production/Operation Management in Virtual Organizations" and "Success Factor Flexibility" as well as author of the books "Enterprise without Boundaries", "Competitive Strategies" and "Web-based Manufacturing". He published numerous articles in books and international journals as well as presentations on international conferences in the fields of production/operations management, information technologies, and business administration. He is member of several scientific and professional organizations, e.g. The

Institute of Electrical and Electronics Engineers (IEEE), the German Association of Engineers (VDI), the German Informatics society (GI), and the Schmalenbach-Gesellschaft fuer Betriebswirtschaft (SG), the oldest association concerned with business administration in Germany. Main research interests: production/operations management, strategic management, industrial information systems, internet-based production systems, mass customization manufacturing systems, virtual organizations, and supply chain management.

email: blecker@ieee.org
Homepage: http://www.manufacturing.de/

Gerhard Friedrich

is professor at the Department of Computer Science and Manufacturing, University of Klagenfurt, Austria. He holds a masters degree in computer sciences (with honors) and a PhD from the Technical University of Vienna, Austria, where he also finished his habilitation in 1994. He was a visitor scientist at the Stanford Research Institute and member of the Office for Object-oriented and Knowledge-based Configuration, Siemens AG, Austria. Gerhard Friedrich was either member of the organization committee or program chair of many international workshops and conferences, e.g. the Austrian Workshop on Model-Based Reasoning '91, the ECAI Workshop on Model-Based Reasoning '92, the German conference on AI 2001 and several workshops on Configuration Systems (AAAI 1999, ECAI 2000, IJCAI 2001, ECAI 2002, IJCAI 2003). Furthermore, he worked in the international research project "CAWICOMS – Customer-Adaptive Web Interface for the Configuration of Products and Services with Multiple Suppliers", is member of the advisory board of the "International Journal of Mass Customization" and member of the board of advisors of Configworks, a software company in the field of personalized handling and servicing of customers via various distribution channels. Gerhard Friedrich is member of several scientific and professional organizations, e.g. The Institute of Electrical and Electronics Engineers (IEEE). He published several edited journals, articles in books and international journals as well as presentations on international conferences in the fields of computer sciences and information technologies. Main research interests: personalization of web-based information systems, configuration systems, mass customization.

email: gerhard.friedrich@ifit.uni-klu.ac.at
Homepage http://www.ifi.uni-klu.ac.at/IWAS/GF/

Bernd Kaluza

is professor at the Department of Production/Operations Management, Business Logistics and Environmental Management, University of Klagenfurt, Austria. He worked several years in industrial practice and holds a masters degree in business administration of the University of Cologne, Germany, as well as a PhD from the University of Mannheim, Germany, where he wrote his habilitation thesis in 1987. Bernd Kaluza was professor for Business Administration, in particular production/operations management, at the Technical University of Karlsruhe and the University of Duisburg. He is co-editor of many books, e.g. "Production/Operation Management in Virtual Organizations" and "Success Factor Flexibility", as well as author of various books, e.g. "Dynamic Product Differentiation Strategy" and "Competitive Strategies". Bernd Kaluza published numerous articles in books and international journals as well as presentations on international conferences in the fields of production/operations management, business logistics, green manufacturing, and business administration. He is member of several scientific and professional organizations, e.g. the German Association for Operations Research (GOR), the European Institute for Advanced Studies in Management (EISAM) and the Schmalenbach-Gesellschaft fuer Betriebswirtschaft (SG), the oldest association concerned with business administration in Germany. Furthermore, he is vice-chairman of the Department of Production/Operations Management of the German Association of Professors in Business Administration (VHB). Main research interests: production/operations management, strategic management, cost management, mass customization, green manufacturing, e-procurement, and supply chain management.

email: bernd.kaluza@uni-klu.ac.at
Homepage: http://www.uni-klu.ac.at/plum/

Gerold Kreutler

is a PhD candidate and research assistant at the Department of Computer Science and Manufacturing at the University of Klagenfurt, Austria within the interdisciplinary multi-year research projects "Modeling, Planning, and Assessment of Business Transformation Processes in the Area of Mass Customization" and "TECTRANS – Technology Transfer". He holds a master degree in computer science (with honors) and is working in the domain of configuration systems for several years, especially in consideration of online customer advisory. Gerold Kreutler participated in several international conferences and workshops in the areas of computer science and business administration. He has taken part in several consultancy projects for the implementation of Enterprise Resource Planning systems. Main research interests: personalization of web-based information systems, business process management and the application of ERP systems.

email: gerold@kreutler.net
Homepage: http://www.kreutler.net/gerold/